Roger Norris and Mike Wooster

Cambridge International AS & A Level

Chemistry

Practical Workbook

CAMBRIDGE UNIVERSITY PRESS

University Printing House, Cambridge CB2 8BS, United Kingdom

One Liberty Plaza, 20th Floor, New York, NY 10006, USA

477 Williamstown Road, Port Melbourne, VIC 3207, Australia

314–321, 3rd Floor, Plot 3, Splendor Forum, Jasola District Centre, New Delhi–110025, India

79 Anson Road, #06–04/06, Singapore 079906

Cambridge University Press is part of the University of Cambridge.

It furthers the University's mission by disseminating knowledge in the pursuit of education, learning and research at the highest international levels of excellence.

www.cambridge.org
Information on this title: www.cambridge.org/9781108539098

© Cambridge University Press 2018

This publication is in copyright. Subject to statutory exception and to the provisions of relevant collective licensing agreements, no reproduction of any part may take place without the written permission of Cambridge University Press.

First published 2018

20 19 18 17 16 15 14 13 12 11 10 9 8 7 6 5 4 3

Printed in Great Britain by CPI Group (UK) Ltd, Croydon CR0 4YY

A catalogue record for this publication is available from the British Library

ISBN 978-1-108-53909-8 Paperback

Cambridge University Press has no responsibility for the persistence or accuracy of URLs for external or third-party internet websites referred to in this publication, and does not guarantee that any content on such websites is, or will remain, accurate or appropriate. Information regarding prices, travel timetables, and other factual information given in this work is correct at the time of first printing but Cambridge University Press does not guarantee the accuracy of such information thereafter.

All examination-style questions, sample mark schemes, solutions and/or comments that appear in this book were written by the author. In examination, the way marks would be awarded to answers like these may be different.

....

NOTICE TO TEACHERS IN THE UK
It is illegal to reproduce any part of this work in material form (including photocopying and electronic storage) except under the following circumstances:
(i) where you are abiding by a licence granted to your school or institution by the Copyright Licensing Agency;
(ii) where no such licence exists, or where you wish to exceed the terms of a licence, and you have gained the written permission of Cambridge University Press;
(iii) where you are allowed to reproduce without permission under the provisions of Chapter 3 of the Copyright, Designs and Patents Act 1988, which covers, for example, the reproduction of short passages within certain types of educational anthology and reproduction for the purposes of setting examination questions.

Contents

Introduction		vii
Safety		viii
AS Practical Skills		ix
A Level Practical Skills		xii

1 Masses, moles and atoms	1.1	Empirical formula of hydrated copper(II) sulfate crystals	2
	1.2	Relative atomic mass of magnesium using molar volumes	4
	1.3	Percentage composition of a mixture of sodium hydrogen carbonate and sodium chloride	7
	1.4	Relative atomic mass of calcium by two different methods: molar volume and titration	9
2 Structure and bonding	2.1	Physical properties of three different types of chemical structure	13
	2.2	Effect of temperature on the volume of a fixed mass of gas	15
	2.3	Effect of pressure on the volume of a fixed mass of gas	17
3 Enthalpy changes	3.1	Enthalpy change for the reaction between zinc and aqueous copper(II) sulfate solution	19
	3.2	Enthalpy change of combustion of alcohols	22
	3.3	Enthalpy change of thermal decomposition	25
	3.4	Change in enthalpy of hydration of copper (II) sulfate	27
4 Redox reactions	4.1	Understanding redox (I): investigating a reactivity series and displacement reactions	30
	4.2	Understanding redox (II): investigating further reactions	32
5 Chemical equilibrium	5.1	Applying Le Chatelier's principle to a gaseous equilibrium	35
	5.2	Applying Le Chatelier's principle to an aqueous equilibrium	37
	5.3	The equilibrium constant for the hydrolysis of ethyl ethanoate	39
6 Rates of reaction	6.1	Effects of concentration on rate of chemical reaction	43
	6.2	Effects of temperature and a homogeneous catalyst on the rate of chemical reaction	46
	6.3	Observing a catalysed reaction	47
7 The properties of metals	7.1	Properties of metal oxides and metal chlorides across Period 3	49
	7.2	Relative atomic mass of magnesium using a back-titration method	51
	7.3	Planning investigation into the separation of two metal ions in solution	53
	7.4	Identification of three metal compounds using qualitative analysis	54

Cambridge International AS & A Level Chemistry

8 **The properties of non-metals**	8.1	Formula of hydrated sodium thiosulfate crystals	56
	8.2	Preparation and properties of the hydrogen halides	58
	8.3	Reaction of bromine with sulfite ions (sulfate (IV))	60
	8.4	Identification of unknowns containing halide ions	62
9 **Hydrocarbons and halogenoalkanes**	9.1	Cracking of hydrocarbons	64
	9.2	How halogenoalkane structure affects the rate of hydrolysis	66
10 **Organic compounds containing oxygen**	10.1	Identifying four unknown organic compounds	68
11 **More about enthalpy changes**	11.1	Enthalpy change of vaporisation of water	72
	11.2	Enthalpy change of solution of chlorides	74
	11.3	Planning investigation into thermal decomposition of iron(II) ethanedioate	77
	11.4	Planning investigation into thermal decomposition of metal carbonates	80
	11.5	Data analysis investigation into enthalpy change of mixing	82
12 **Electrochemistry**	12.1	Determining the Faraday constant	86
	12.2	Comparing the voltage of electrochemical cells	88
	12.3	Half-cells containing only ions as reactants	90
	12.4	Planning investigation into changing the concentration of ions in an electrochemical cell	91
	12.5	Planning and data analysis investigation into electrical conductivity of ethanoic aid	93
13 **Further aspects of equilibria**	13.1	Change in pH during an acid–base titration	95
	13.2	Data analysis investigation into partition of ammonia between water and trichloromethane	97
	13.3	Planning investigation into an esterification reaction at equilibrium	99
	13.4	Planning investigation into the effect of temperature on the $N_2O_4 \rightleftharpoons 2NO_2$ equilibrium	101
	13.5	Data analysis investigation into equilibrium entropy and enthalpy change	104
14 **Reaction kinetics**	14.1	Kinetics of the reaction between propanone and iodine	107
	14.2	Data analysis investigation into rate of decomposition of an organic compound	109
	14.3	Planning investigation into determination of the order of a reaction	111
	14.4	Planning investigation into the effect of temperature on rate of reaction	114
15 **Transition elements**	15.1	Planning investigation into copper content of copper ore	118
	15.2	Data analysis investigation of iron tablets	120
	15.3	Data analysis investigation into formula of a complex ion	122
	15.4	Planning investigation into reaction of copper with potassium dichromate(VI)	124

16 More about organic chemistry	16.1	Planning investigation into making an azo dye	127
	16.2	Data analysis investigation into acylation of a nucleic acid	129
	16.3	Planning investigation into nitration of benzene	131
17 Identifying organic compounds	17.1	Data analysis investigation into extracting an amino acid from hair	133
	17.2	Data analysis investigation into identification of a white crystalline solid	134
	17.3	Data analysis investigation into preparation and identification of a colourless liquid	136

Introduction

Practical work is an essential part of any advanced Chemistry course. For Cambridge International AS & A Level Chemistry, Paper 3 and Paper 5 focus on the assessment of practical skills.

The practical investigations in the workbook have been carefully chosen to:

- meet the requirements of all the learning objectives for specific practical activities
- provide progressive guidance and practice of Assessment Objective 3 (AO3) skills.

The skills grids at the front of this guide summarise the practical skills that are assessed in Paper 3 (AS) and Paper 5 (A level). You can use these grids to search for practical investigations that involve a particular skill or skills. At the beginning of each practical investigation, the learning objectives and skills that are supported are also listed.

The order of the investigations presented follows the order of the topics in the Cambridge International AS & A Level Chemistry coursebook, but please note that this does not mean that they must be completed in that order. Some coursebook chapters involve the use of quantitative techniques. All techniques listed in the practical guidance are covered in the workbook.

⚙ These points have been provided to give extra support to students who may be struggling with the investigation.

Each chapter of the workbook has more than one investigation so do choose those that you feel suits the equipment and time that you have available. Chemicals required by the investigations in the workbook have been chosen to be as commonly available as possible and virtually all the equipment required is that listed in the practical guidance. We have, however, provided a set of sample results for each Practical Investigation, which you can give to learners who have not managed to obtain a complete set of results themselves, so that they can continue to answer all of the Data analysis and Evaluation questions.

Although practical work requires time, it is time well spent. Practical work enables learners to acquire transferrable skills and gives them the confidence that the theory they have learned works in practice. Because of this, the details of the theory are more easily retained. The important learning experiences, when carrying out practical work, are the range of skills that are being used and developed – the *processes* of planning, carrying out, observing, recording, and analysing. The workbook gives the learners experience in developing these skills. It is not a series of mock practical exam papers! But, in carrying out the investigations, the learners will practise and acquire the skills that will enable them to be more confident when tackling their practical exam.

⚙ These points provide additional tasks to extend more able learners.

Safety

Working safely in a science laboratory is an essential aspect of learning which characterises practical work. It is the duty of the school to make it clear to learners just what is expected of them when they are working in a laboratory.

In every investigation, every learner is expected to wear eye protection and long hair should be tied back. Safety goggles generally give more protection than safety spectacles. It is also advisable for them to wear a laboratory coat to protect their clothing from chemical splashes.

All chemicals should be treated as hazardous and whenever they are spilt on the skin they should be washed off immediately using water. The apparent dangers of a chemical may not have been realised and therefore using them without safety precautions can lead to unforeseen problems.

Learners should also take responsibility for working safely. It is advisable that learners are made aware of the hazard signs on reagent bottles and in the laboratory. Table S1. shows some of the most common hazard symbols. This is not an exhaustive list, but it does list the most common hazard symbols encountered in school science laboratories. An up-to-date list of CLEAPPS hazards is available for students to download.

Hazard symbol	What does it mean?	Special points
	The substance is **corrosive**. It will damage your skin and tissues if it comes into contact with them.	Always wear safety goggles and if possible gloves when using corrosive substances.
	The substance is an **irritant**. If it comes into contact with your skin it can cause blisters and redness.	Always wear safety spectacles when using irritants.
	The substance is **toxic** and can cause death if swallowed, breathed in or absorbed by skin.	Wear gloves and eye protection.
	The substance is **flammable** and catches fire easily.	Keep the substance away from naked flames and when heating reaction mixtures use the hot water from a kettle rather than using Bunsen burners.
	The material is a **biohazard**. Examples are bacteria and fungi.	Seek advice about specific biohazards.
	The substance is an **oxidising agent**. It will liberate oxygen when heated or in the presence of a suitable catalyst.	Keep oxidising agents well away from flammable materials.

Table S1

AS Practical Skills

The following grids map the practical investigations from the workbook to the mark categories for Papers 3 and 5, as listed in the Cambridge International AS & A Level Chemistry syllabus.

The grids are designed to aid you when planning practical and theory lessons, to ensure learners develop the practical skills required as part of this course.

Manipulation, measurement and observation (MMO)

SKILL	CHAPTER									
	1	2	3	4	5	6	7	8	9	10
Collection of data and observations										
(a) set up apparatus correctly	1.1; 1.2; 1.3; 1.4	2.1; 2.2; 2.3	3.1; 3.2; 3.3; 3.4		5.1; 5.2; 5.3	6.1;	7.2; 7.3; 7.4	8.1; 8.2	9.1; 9.2	10.1
(b) follow instructions in written form or from diagrams	1.1; 1.2; 1.3; 1.4	2.1; 2.2; 2.3	3.1; 3.2; 3.3; 3.4	4.1; 4.2	5.1; 5.2; 5.3	6.1; 6.2	7.1; 7.2; 7.3; 7.4	8.1; 8.3; 8.4	9.1; 9.2	10.1
(c) use apparatus to collect an appropriate quantity and quality of data and observations	1.1; 1.2; 1.3; 1.4	2.1; 2.2; 2.3	3.1; 3.2; 3.3; 3.4	4.1; 4.2	5.1; 5.2; 5.3	6.1; 6.2	7.1; 7.2; 7.3; 7.4	8.1; 8.2; 8.3: 8.4	9.1; 9.2	10.1
(d) make measurements using pipettes, burettes, and other common laboratory apparatus	1.1; 1.2; 1.3; 1.4	2.1; 2.2; 2.3	3.1; 3.2; 3.3; 3.4		5.3	6.1	7.2	8.1		
(e) make accurate and consistent measurements and observations	1.1; 1.2; 1.3; 1.4	2.1; 2.2; 2.3	3.1; 3.2; 3.3; 3.4	4.1; 4.2	5.2; 5.3	6.1; 6.2; 6.3	7.1; 7.2; 7.3; 7.4	8.1; 8.2; 8.3; 8.4	9.1; 9.2	10.1
Decisions relating to measurements or observations										
(a) decide how many tests or observations to perform	1.1; 1.3; 1.4	2.2		4.1; 4.2	5.1; 5.3		7.2;	8.1; 8.2; 8.3	9.1	10.1
(b) make a suitable range and number of measurements suitable for the experiment	1.1; 1.3; 1.4	2.2	3.2; 3.4	4.2	5.3		7.2;	8.1; 8.3	9.1	
(c) decide how long to leave experiments running before making readings	1.1; 1.2; 1.4	2.1; 2.2	3.4	4.2	5.1; 5.2; 5.3	6.1	7.2; 7.3; 7.4	8.2; 8.3; 8.4	9.1; 9.2	10.1
(d) make as many repeated readings or observations as appropriate	1.1; 1.2;1.3; 1.4	2.1; 2.2	3.3	4.1	5.1; 5.2		7.1; 7.2	8.1; 8.2	9.1	10.1
(e) identify where confirmatory tests are needed and the nature of these tests	1.3			4.1; 4.2						10.1
(f) choose reagents to distinguish between given ions				4.1; 4.2			7.3; 7.4	8.3	9.2	

Presentation of data and observations (PDO)

SKILL	CHAPTER									
	1	2	3	4	5	6	7	8	9	10
Recording data and observations										
(a) present data, values or observations in a single table of results	1.2; 1.3	2.1; 2.2	3.1; 3.2; 3.3; 3.4	4.2	5.1; 5.2; 5.3	6.1;	7.2;	8.1	9.1; 9.2	10.1

SKILL	CHAPTER									
	1	2	3	4	5	6	7	8	9	10
Recording data and observations (Continued)										
(b) draw up a table in advance of taking readings				4.2				8.1	9.1	10.1
(c) include in the results table: columns for raw data, calculated values and analyses	1.3; 1.4	2.1; 2.2	3.3; 3.4	4.2		6.1; 6.2	7.2	8.1		10.1
(d) use column headings containing both quantities and suitable scientific units	1.4		3.3; 3.4			6.2		8.1	9.1	
(e) record readings to the same degree of precision and observations to the same level of detail	1.1; 1.2; 1.3; 1.4	2.1; 2.2	3.1; 3.3; 3.4		5.1; 5.2; 5.3	6.1;	7.1; 7.2	8.1; 8.3; 8.4	9.1	
Display of calculations and reasoning										
(a) show all working in calculations and the key steps in your reasoning	1.1; 1.2; 1.3; 1.4	2.2; 2.3	3.1; 3.2; 3.3; 3.4		5.3	6.1	7.2	8.1		
(b) use the correct number of significant figures for calculated quantities	1.1; 1.2; 1.3; 1.4	2.2; 2.3	3.1; 3.2; 3.3; 3.4		5.3	6.1	7.2	8.1		
Data layout										
(a) choose a suitable, clear way of presenting data, for example, tables, graphs or a mixture			3.1; 3.4	4.2		6.1	7.1	8.1	9.1; 9.2	10.1
(b) decide how to plot the variables and whether a graph should be a straight line or a curve	1.1; 1.2	2.2	3.1; 3.4			6.1				
(c) plot appropriate variables on clearly labelled *x*- and *y*-axes	1.1; 1.2	2.2	3.1; 3.4			6.1				
(d) choose suitable scales for graph axes	1.1; 1.2	2.2	3.1; 3.4			6.1				
(e) plot all points or bars to an appropriate accuracy	1.1; 1.2	2.2	3.1; 3.4			6.1				
(f) draw best-fit lines taking into account the anomalous points	1.1; 1.2	2.2	3.1; 3.4			6.1;				

Analysis, conclusions and evaluation (ACE)

SKILL	CHAPTER									
	1	2	3	4	5	6	7	8	9	10
Data interpretation and sources of error										
(a) describe the patterns and trends shown by tables and graphs		2.2; 2.3	3.1; 3.4		5.1; 5.2	6.1; 6.2	7.1		9.2	
(b) describe and summarise the key points of a set of observations		2.1; 2.2; 2.3			5.1; 5.2	6.1	7.1		9.2	

AS Practical Skills

SKILL	CHAPTER									
	1	2	3	4	5	6	7	8	9	10
Data interpretation and sources of error (Continued)										
(c) find an unknown value from a graph including the drawing of intersecting points	1.1; 1.2	2.2	3.1; 3.4							
(d) calculate other quantities from data or the mean from reproducible values	1.1; 1.2; 1.3;	2.2; 2.3	3.4		5.3	6.1	7.2	8.1	9.2	
(e) determine the gradient of a straight-line graph						6.1				
(f) evaluate the effectiveness of control variables										
(g) identify the most significant sources of error in an experiment	1.3; 1.4	2.2	3.1; 3.2 3.3		5.3		7.2			
(h) estimate in terms of number values the uncertainty in quantitative measurements	1.2; 1.3; 1.4		3.1; 3.2; 3.3		5.3					
(i) express these uncertainties as an actual or percentage error										
(j) show and understand the difference between systematic and random errors	1.2; 1.3; 1.4		3.1; 3.2							
Drawing conclusions										
(a) consider to what extent the experimental data support a given hypothesis	1.2; 1.3	2.3		4.2	5.1; 5.2; 5.3	6.1; 6.2; 6.3	7.1	8.3	9.2	10.1
(b) make further predictions arising from the experiment				4.2	5.1; 5.2	6.1; 6.3				
(c) draw conclusions from observations, data and calculated values	1.1; 1.2; 1.3; 1.4	2.1; 2.2; 2.3	3.1; 3.2; 3.3	4.1; 4.2	5.1; 5.3	6.1; 6.2; 6.3	7.1; 7.2	8.1; 8.2; 8.3; 8.4	9.1; 9.2	10.1
(d) make scientific explanations arising from the data, observations and conclusions	1.3; 1.4	2.1; 2.2		4.1; 4.2	5.1; 5.3	6.1; 6.2; 6.3	7.1	8.2; 8.3; 8.4	9.1; 9.2	10.1
Suggesting improvements										
(a) suggest modifications that improve the accuracy of the experiment/observations	1.1; 1.3; 1.4					6.1; 6.2				
(b) suggest how to extend the investigation to answer a new question						6.2	7.3			
(c) describe modifications to the experiment in words or diagrams	1.4					6.2				

A Level Practical Skills

Planning an investigation (PI)

SKILL	CHAPTER						
	11	12	13	14	15	16	17
Selecting information							
(a) locate, select, organise and present information from a variety of sources				14.3			17.1; 17.2; 173
(b) construct arguments to support hypotheses and justify a course of action	11.1; 11.3; 11.5			14.3; 14.4	15.2	16.1; 16.3	17.1; 17.2; 17.3
(c) apply knowledge, including principles, to new situations	11.1; 11.3; 11.5	12.3; 12.5	13.4; 13.5	14.1; 14.2; 14.4	15.1; 15.3; 15.4	16.1; 16.2; 16.3	17.1; 17.2; 17.3
Defining the problem under investigation							
(a) identify a safe, efficient procedure that leads to a reliable result	11.3; 11.4,	12.5	13.3; 13.4	14.3; 14.4	15.1; 15,4	16.1; 16.3	
(b) express the aim in terms of a prediction (in words or as a predicted graph)				14.3; 14.4			
(c) identify the steps needed to carry out the procedure	11.3; 11.4	12.4; 12.5	13.3; 13.4	14.3; 14.4	15.1; 15.4	16.1; 16.3	
(d) identify apparatus that is suitable for carrying out each step of the procedure	11.3; 11.4	12.3; 12.4; 12.5	13.3; 13.4	14.3; 14.4	15.1; 15.4	16.1; 16.3	17.1
(e) indicate how and why the procedure suggested will be effective			13.2; 13.4	14.4	15.3	16.1	
Control experiments and identification of variables							
(a) identify the independent variable and the dependent variable in an experiment or investigation	11.5	12.4		14.4; 14.4			
(b) explain how control experiments verify that no other factors influence the variables		12.5			15.3; 15.4	16.1; 16.2	
(c) identify any variables that are to be controlled	11.1; 11.3; 11.4	12.1; 12.5	13.2; 13.3	14.1; 14.3; 14.4	15.1; 15.3; 15.4	16.2	
Considering hazards							
(a) assess the risks of the proposed experiment	11.1; 11.3; 11.4; 11.5	12.1; 12.5	13.3; 13.4	14.3; 14.4	15.1; 15.4	16.1; 16.3	
(b) describe precautions that should be taken to keep risks to a minimum	11.4	12.5	13.2; 13.3; 13.4	14.3; 14.4	15.1; 15.4	16.1; 16.3	

Carrying out an investigation (COI)

SKILL	CHAPTER						
	11	12	13	14	15	16	17
Methods used							
(a) describe the method to be used when carrying out an investigation	11.1; 11.3; 11.4	12.3; 12.4; 12.5	13.3; 13.4	14.3; 14.4	15.1; 15.4	16.1	17.1
(b) describe the arrangement of the apparatus and the steps in the procedure to be followed	11.3; 11.4	12.4; 12.5	13.3; 13.4	14.3; 14.4	15.1; 15.4	16.1	17.1

A Level Practical Skills

SKILL	CHAPTER						
	11	12	13	14	15	16	17
Methods used (Continued)							
(c) arrange and use the apparatus provided correctly	11.1; 11.2; 11.4	12.1; 12.2; 12.3; 12.4	13.1	14.3; 14.4			
(d) suggest and use appropriate volumes and concentrations of reagents	11.3; 11.4	12.3; 12.4; 12.5	13.3; 13.4	14.3; 14.4	15.4	16.1; 16.3	
Carrying out the experiment							
(a) carry out the experiment by varying the independent variable and measuring the dependent variable	11.1; 11.2; 11.4	12.1; 12.2; 12.3; 12.4	13.1	14.3; 14.4			
(b) carry out the experiment so that key variables are controlled	11.1; 11.2; 11.4	12.1; 12.2; 12.3; 12.4	13.1	14.3; 14.4			
(c) carry out the experiment with the required degree of accuracy	11.1; 11.2; 11.4	12.1; 12.2; 12.3; 12.4	13.1	14.3; 14.4			
(d) carry out the experiment safely	11.1; 11.2; 11.4	12.1; 12.2; 12.3; 12.4	13.1	14.3; 14.4			

Handling information (HI)

SKILL	CHAPTER						
	11	12	13	14	15	16	17
Collecting and displaying data							
(a) describe the outcome of steps in the procedure relevant to the experiment	11.3; 11.4			14.3; 14.4			
(b) handle information, distinguishing the relevant from the irrelevant	11.1; 11.2; 11.3; 11.5	12.1; 12.3; 12.4; 12.5	13.1; 13.2; 13.3; 13.4	14.1; 14.2; 14.3; 14.4	15.2; 15.3		17.1; 17.2; 17.3
(c) draw up tables for data that need to be recorded	11.1; 11.2	12.4	13.1	14.3; 14.4			
Manipulating data							
(a) describe how the data might be used in order to reach a conclusion			13.1	14.1; 14.3; 14.4	15.2		17.2; 17.3
(b) manipulate numerical and other data	11.1; 11.2; 11.4; 11.5	12.1; 12.2; 12.4; 12.5	13.1; 13.2; 13.3; 13.5	14.1; 14.2; 14.3; 14.4	15.1; 15.2; 15.3; 15.4	16.2	17.2; 17.3
(c) translate information from one form to another including graphical information	11.1; 11.2; 11.5	12.4; 12.5	13.1; 13.2; 13.5	14.2; 14.3; 14.4	15.1; 15.3	16.2	

Data analysis (DA)

SKILL	CHAPTER						
	11	12	13	14	15	16	17
Identifying trends and patterns							
(a) analyse information so as to identify patterns and report trends	11.2; 11.5	12.1; 12.2; 12.4; 12.5	13.1; 13.2; 13,3	14.1; 14.2; 14.3; 14.4	15.3; 15.4	16.2	17.1; 17.2; 17.3
(b) use tables and graphs of quantitative data to draw attention to key points	11.1			14.3; 14.4	15.4	16.2	17.2; 17.3
(c) comment, where necessary on the variability of the data	11.5	12.4	13.3	14.2; 14.3; 14.4	15.2; 15.3		
(d) analyse data from spectra or other published data to reach appropriate conclusions					15.3; 15.4		17.2; 17.3
Identifying and using calculations							
(a) identify calculations required and the means of presentation of data provided	11.3	12.1	13.1		15.1; 15.2		
(b) AL3.2.3 use calculations to enable simplification or explanation of data	11.1; 11.3	12.4	13.3; 13.5	14.1; 14.2	15.2; 15.3		

Conclusions and Predictions (CP)

SKILL	CHAPTER						
	11	12	13	14	15	16	17
Drawing conclusions							
(a) analyse qualitative data to draw appropriate conclusions	11.2; 11.4		13.1		15.1; 15.4		
(b) analyse quantitative data provided to draw conclusions	11.3; 11.4; 11.5	12.3; 12.4; 12.5	13.1; 13.2; 13.5	14.1; 14.3; 14.4	15.2; 15.3	16.2	17.1; 17.2; 17.3
(c) draw conclusions to describe the key features of the data and analyses			13.1	14.1; 14.3; 14.4			17.2; 17.3
(d) make detailed explanations of the data, analyses and conclusions				14.1; 14.3			17.2; 17.3
(e) consider whether the experimental data supports the conclusion reached	11.2; 11.3	12.2	13.3	14.3; 14.4	15.3		17.1; 17.2; 17.3
Making predictions							
(a) make further predictions, ask relevant questions and suggest improvements	11.2	12.1; 12.4	13.4	14.4			
(b) suggest improvements by asking relevant questions	11.2	12.1; 12.4	13.4	14.4			

A Level Practical Skills

Evaluating investigations (EI)

SKILL	CHAPTER						
	11	12	13	14	15	16	17
Identifying problems with the procedure							
(a) identify and explain the weaknesses of the experimental procedure used	11.1; 11.2; 11.4; 11.5	12.1; 12.2; 12.4	13.1; 13.3; 13.4	14.1,14.2; 14.3,14.4	15.1; 15.2		
(b) explain the effect of the incorrect use of apparatus on the results		12.4		14.1	15.3		
(c) use information provided to assess the effectiveness of the control of the variables				14.2; 14.3; 14.4		16.2	
(d) explain how changes in the conditions used may affect the results		12.1		14.3; 14.4			
(e) explain how changes in concentration of reagents may affect the results		12.2; 12.4		14.3			
Identifying problems with the data							
(a) identify anomalous values in data provided, give possible explanations and suggest how to deal with these.	11.5	12.5	13.1; 13.5		15.3		
(b) identify the extent to which readings provided have been reproduced	11.5		13.3		15.2		
(c) describe whether the range of data provided is sufficient	11.5		13.2; 13.5	14.2; 14.3; 14.4	15.2; 15.3		
Making a judgement on the conclusions							
(a) evaluate information and hypotheses	11.1; 11.2; 11.3; 11.4	12.1; 12.3; 12.4; 12.5	13.1; 13.2; 13.3; 13.4	14.1; 14.2; 14.3; 14.4	15.2; 15.3; 15.4		17.1
(b) evaluate information to make judgements on the confidence of the conclusions drawn	11.1; 11.2; 11.4; 11.5			14.1;14.2; 14.4	15.1; 15.3	16.2	17.1

Masses, moles and atoms

Chapter outline

This chapter relates to Chapter 1: Moles and equations, Chapter 2: Atomic structure and Chapter 3: Electrons in atoms in the coursebook.

In this chapter learners will complete practical investigations on:

- 1.1 Empirical formula of hydrated copper(II) sulfate crystals
- 1.2 Relative atomic mass of magnesium using molar volumes
- 1.3 Percentage composition of a mixture of sodium hydrogen carbonate and sodium chloride
- 1.4 Relative atomic mass of calcium by two different methods: molar volume and titration

Practical investigation 1.1:
Empirical formula of hydrated copper(II) sulfate crystals

Introduction

In this investigation learners determine the empirical formula (see Chapter 1 of the coursebook) of hydrated copper(II) sulfate by finding the value of **x** in $CuSO_4.xH_2O$. They weigh out some hydrated copper(II) sulfate in an evaporating basin, heat it to constant mass, determine the mass of water present in their sample and then find the molar ratio: $CuSO_4 : H_2O$.

Skills focus

The following skill areas are developed and practised (see the skills grids at the front of this guide for codes):

MMO	Collection of data and observations: (a), (b), (c), (d) and (e)
	Decisions relating to measurements of observations: (a), (b) (c) and (d)
PDO	Recording data and observations: (e)
	Display of calculations and reasoning: (a) and (b)
	Data layout: (b), (c), (d), (e) and (f)
ACE	Data interpretation and sources of error: (c) and (d)
	Drawing conclusions: (c)
	Suggesting improvements: (a)

Duration

This investigation should take no more than 1 h to complete. However, as it is the first time learners will have completed error calculations you may need another hour to go through the errors involved.

Preparing for the investigation

- Learners should be made aware of the 'Skills Chapter' and how it informs them about the techniques they will be using.
- They will also need to have an awareness of the sources of errors.
- Learners will need to understand the concept of an empirical formula and be able to calculate the number of moles present.
- They should revise the concepts of moles and molar ratios.

Equipment

Each learner or group will need:

- a pipe-clay triangle
- an evaporating basin
- Bunsen burner and tripod
- tongs
- glass stirring rod
- two heat-resistant pads
- spatula

Access to:

- a supply of gas
- a top-pan balance that reads to at least two decimal places

Chapter 1: Masses, moles and atoms

Safety considerations

- Learners must wear eye protection at all times in this experiment and tie back long hair.
- When weighing the evaporating basin and copper sulfate the learners should place it on the extra heat-resistant mat and then carry it across to the top-pan balance.
- The copper(II) sulfate is an environmental hazard and should be recycled. It can be used as a test for water or dissolved in water and recrystallised. It could also be used in a Hess' Law determination.

Carrying out the investigation

- They may need help to understand what is meant by 'water of crystallisation' and how it is loosely bound to the copper(II) sulfate and that the number of water molecules per formula is a whole number.
- Assuming that the length of the practical time available is about 1 h then this is sufficient time for each group to do one determination.
- Allocate a given mass to each group. It is a good idea to give the larger masses of copper sulfate to the more able learners or more patient ones because they will obviously need more time in heating the copper(II) sulfate to give the anhydrous form.
- If they heat the copper sulfate properly there will be some at the beginning that will stick to the stirring rod and the basin and when this ceases to happen it shows that they are removing the water from the salt.
- The anhydrous salt should be as near white as possible but may have a greyish tinge after the heating is finished and constant mass is obtained.
- Ensure that if more than one balance is used, the learners should use the same balance throughout. By doing this any errors in the balances are reproducible.
- Some learners will need help on why some points on their graph lie above and below the line.
- Some will also need help on heating the copper(II) sulfate as gently as possible (see above) so will need to be trained on how to adjust the Bunsen flame to a very low level.
- Learners who struggle with the practical, especially the theoretical part, should be given the lowest value masses so that their heating is over quickly and they can start processing their results.
- More able learners should, if possible, be allowed to work on their own.

Common learner misconceptions

- When instructed to 'heat gently' some learners will still use a yellow flame.

Sample results

Mass of crystals/g	Mass of anhydrous copper(II) sulfate/g
0.20	0.12
0.50	0.32
0.80	0.51
2.50	1.60

Table 1.1

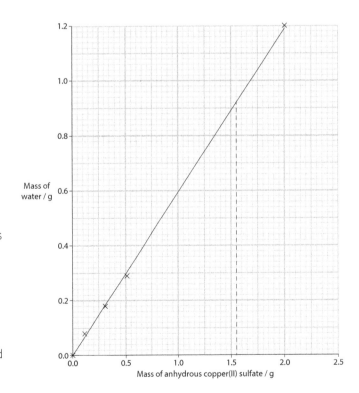

Figure 1.1

Cambridge International AS & A Level Chemistry

> **Answers to the workbook questions (using the sample results)**
>
> **a & b** It is quite easy to get a set of results that give the ideal answer for this practical (see Table 1.1 and Figure 1.1).
>
> **c** As can be seen from the graph, the mass of water that combines with 1.60 g of copper(II) sulfate is 0.90 g
>
> **d** Results shown in Table 1.2
>
> | | Copper(II) sulfate ($CuSO_4$) | Water (H_2O) |
> |---|---|---|
> | Mass/g | 1.60 | 0.90 |
> | Number of moles | $\frac{1.60}{159.6} = 0.0100$ | $\frac{0.90}{18} = 0.0500$ |
> | Simplest ratio (divide by lowest number) | $\frac{0.0100}{0.0100} = 1$ | $\frac{0.0500}{0.0100} = 5$ |
>
> Table 1.2
>
> **e** This means that the formula for the hydrated copper(II) sulfate is $CuSO_4 \cdot 5H_2O$
>
> **f** 0, 0 because if there is no copper(II) sulfate then there will be no water attached to it.
>
> **g i** If a point lies **above the line** then it could have been heated too much and the copper (II) sulphate has decomposed to some extent.
>
> **ii** If a point lies **below the line** there has been insufficient heating of the crystals and the water of crystallisation is still attached to them. However, it could be that the heated solid has been left to cool and absorbed water from the atmosphere.
>
> **h** The best alternative is to use an oven. The temperature of the oven can be adjusted to one where the water of crystallisation will be removed but it will not be hot enough to decompose the copper(II) sulfate. Using a Bunsen burner cannot be sufficiently accurate. A possible way of determining the Bunsen burner temperature is to use a thermocouple to give a reading of the temperature. Even using this method is inaccurate because any slight change in the extent to which the air hole is opened will lead to a change in temperature.

Practical Investigation 1.2:
Relative atomic mass of magnesium using molar volumes

Skills focus

The following skill areas are developed and practised (see the skills grids at the front of this guide for codes):

MMO	Collection of data and observations (a), (b), (c), (d) and (e)
	Decisions relating to measurements of observations (c) and (d)
PDO	Recording data and observations (a) and (e)
	Display of calculations and reasoning (a) and (b)
	Data layout (b), (c), (d), (e) and (f)
ACE	Data interpretation and sources of error (c), (d), (h), (i) and (j)
	Drawing conclusions (a) and (c)

Duration

This investigation should take approximately 1.5 h to complete.

Preparing for the investigation

- Learners should, ideally, have a good understanding of moles and molar volumes. The crucial relationships are:

$$A_r = \frac{\text{mass } (m)}{\text{number of moles } (n)} \text{ and } n = \frac{\text{Volume of gas in cm}^3}{24\,000}$$

Equipment

Each learner or group will need:

- either a trough;
  a selection of measuring cylinders (10 cm³; 25 cm³ and 50 cm³); OR a 100 cm³ gas syringe
- 150 cm³ conical flask with retort stand, boss and clamp
- small piece of steel wool
- 25 cm³ measuring cylinder for acid
- one 10.0 cm length of magnesium ribbon
- 30 cm ruler
- plastic gloves (see safety considerations)

Chapter 1: Masses, moles and atoms

Access to:
- a top-pan balance reading to **at least** two decimal places
- 2 mol dm^{-3} hydrochloric acid

Alternative equipment

- Of the two sets of apparatus suggested, the easiest to set up is the one using the gas syringe. However, if gas syringes are not available, then the displacement of water in a measuring cylinder works very well.

Safety considerations

- Learners must wear eye protection at all times and tie hair back if it is long.
- Magnesium is highly flammable.
- Hydrogen is a flammable gas.
- 2 mol dm^{-3} hydrochloric acid is an irritant.
- Steel wool sometimes splinters and some learners could be quite sensitive to this. To lower the risk plastic gloves should be worn when using the steel wool to clean the magnesium.

Carrying out the investigation

- The point of weighing out 10 cm lengths of magnesium ribbon is that 10 cm will give a valid reading on the top-pan balance, especially if the balance reads to only two decimal places. The masses of the shorter lengths are then calculated using the relationship:
  $$\text{mass} = \frac{\text{length}}{10} \times \text{mass of 10 cm length}$$
- Please note that if learners are measuring the gas volume by displacement of water, the first problem to overcome is making sure that the measuring cylinder is full of water when it is put in the trough and that none or very little escapes. This can be done by either learners holding their hands over the end of the measuring cylinder or placing a piece of plastic wrap over the open end and then turn the measuring cylinder upside down when it is in the trough. Remember to remove the film before starting the actual measurement. A boiling tube will do as well as a conical flask for the reaction vessel.

- The main problem with the practical is the purity of the magnesium ribbon. If you have fresh ribbon then omit the cleaning. If it is visibly oxidised then it will need cleaning and that is done using the steel wool. This should be done by holding the ribbon using the wool and then drawing it through. Once should be enough. Any more than that will lead to irregularities in the thickness of the ribbon and inaccuracy when estimating the masses of the individual lengths.

- Evaluation of a practical method always presents problems to learners and they will need help when estimating the percentage error due to using different apparatus.

- Before the practical, a short demonstration will give learners some idea of the volumes of gas that they will be dealing with. This can be their trial run but more able learners can be asked to do this for themselves. If the volume of gas for a 1 cm length of ribbon is found then they should be able to estimate the volumes for the other lengths and adjust their choice of measuring cylinder (if these are used) accordingly.

- If learners are measuring the gas volume by displacement of water then they can be marked on which measuring cylinder they use for the most accurate measurements of gas volumes.

- Learners can be asked to analyse their results in Microsoft Excel or a similar data-handling application.

Sample results

Mass of 10 cm length of magnesium ribbon = 0.160 g

The results from one set of measurements are shown in Table 1.3

Length of Mg ribbon/cm	Mass of Mg/g	Expt. 1	Expt. 2	Average
0.00	0	0	0	0
0.50	0.008	8	8	8.0
1.00	0.016	16	17	16.5
1.50	0.024	23	25	24.0
2.00	0/032	30	31	30.5

Table 1.3

Answers to the workbook questions (using the sample results)

a Please see Figure 1.2

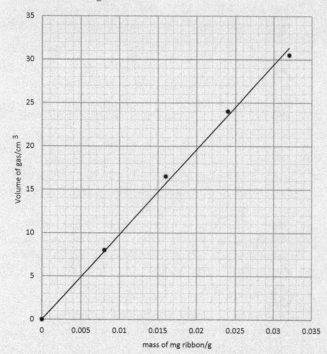

Figure 1.2

b Using Figure 1.2: 24.0 cm³ of H_2 is produced from 0.0245 g of magnesium

c 24.0 cm³ = $\frac{24}{24,000}$ mol = 0.001 mol of H_2 = number of mol of magnesium

Therefore, mass of 1 mol

$= \frac{m}{n} = \frac{0.0245}{0.001} = 24.5 \text{ g mol}^{-1}$

d Percentage error =

$\frac{|\text{Actual value} - \text{experimental value}|}{\text{Actual value}} \times 100$

$= \frac{24.5 - 24.3}{24.3} \times 100\% = 0.823\%$

The mass of 10 cm of magnesium ribbon is around 0.15–0.17 g.

In this experiment, the systematic errors come from the top-pan balance, the ruler and the measuring cylinder (or gas syringe).

e Maximum error from the top-pan balance

If the top-pan balance reads to 0.01 g then the maximum error can be estimated to be ±0.005 g. If we think our 10 cm length of magnesium will weigh in the region of 0.15 g then the percentage

error = $2 \times \frac{0.005}{0.15} \times 100\% = 6.67\%$

f Percentage error from measurements of lengths

For example, if the length is 1 cm then the maximum percentage error is equal to $\frac{0.05}{1.0} \times 100\% = 5\%$

g Total error from length measurements

i The measurement of the lengths of magnesium ribbon. If we go by the rules that the maximum error or uncertainty is half the smallest possible measurement then the ruler reads to ±0.5 mm. The length measurements will give the greatest error.

ii If they use measuring cylinders, learners should be marked on their choice. For example, if they estimate from their trial run that they will obtain around 20 cm³ from the reaction, then choosing a 50 cm³ measuring cylinder that is graduated in 2.0 cm³ divisions will give a maximum error of ±1.0 cm³ (half the graduation's reading).

iii Total possible percentage error from apparatus readings. In this case, the maximum percentage error is $\frac{1}{20} \times 100\% = 5\%$. This error is halved if a 25 cm³ measuring cylinder is used.

h Other factors that limit accuracy and contribute to the error

- Because of the cleaning by the steel wool, the thickness of the magnesium ribbon is not the same along its whole length.

- There may still be some oxide present even after cleaning.

Practical Investigation 1.3:
Percentage composition of a mixture of sodium hydrogen carbonate and sodium chloride

Introduction
In this investigation, learners will find the percentage composition of a mixture of sodium hydrogen carbonate and sodium chloride. They will do this by titrating the sodium hydrogen carbonate against standard hydrochloric acid.

Skills focus
The following skill areas are developed and practised (see the skills grids at the front of this guide for codes):

MMO	Collection of data and observations (a), (b), (c), (d) and (e)
	Decisions relating to measurements of observations: (a), (b) (d) and (e)
PDO	Recording data and observations (a), (c) and (e)
	Display of calculations and reasoning (a) and (b)
ACE	Data interpretation and sources of error (d)
	Drawing conclusions (a), (c) and (d)
	Suggesting improvements (a)

Duration
This investigation requires 1 h of preparation, including making up the solution of the mixture, then 1 h for the titrations and calculations.

Preparing for the investigation
- The volume of hydrochloric acid required can be calculated from the projected titre values. For example, if you calculate that the sodium hydrogen carbonate would require 17.00 cm^3 of acid for complete reaction, then if each learner or group does five titrations, 85 cm^3 is required and 100 cm^3 per learner/group would be an adequate allocation.

Equipment
Each learner or group will need:
- 150 cm^3 conical flask
- 250 cm^3 volumetric flask
- wash bottle of distilled water
- burette stand
- 25.0 cm^3 pipette
- white tile
- 250 cm^3 beaker and 100 cm^3 beaker
- stirring rod and small dropper
- small filter funnel for burette and larger one for volumetric flask
- weighing boat
- 50 cm^3 burette

Access to:
- a mixture of sodium hydrogen carbonate and sodium chloride. You can decide on the composition. If different classes are doing the same practical they can be given different mixtures to investigate.
- 0.100 mol dm^{-3} hydrochloric acid
- The volume of hydrochloric acid required can be calculated from the projected titre values. For example, calculations might show that the sodium hydrogen carbonate would require 17.00 cm^3 of acid for a complete reaction. Therefore, if each student or group does five titrations, 85 cm^3 is required and 100 cm^3 per student/group would be an adequate allocation.
- methyl orange indicator and dropper
- either a two or three place top-pan balance
- distilled water

Safety considerations
- They must wear eye protection and tie hair back if it is long.
- The acid is an irritant at the concentration used in the experiment.
- Methyl orange is poisonous. Wash off skin immediately.

Carrying out the investigation
- As far as the mixture is concerned, a typical calculation is as follows:

    i   Let us suppose we want the titre to be 17.20 cm^3. Such a volume requires the learner to fill up their burette twice at most.

    ii The number of moles of sodium hydrogen carbonate present in 25.00 cm^3 is: 17.20 × 10^{-3} × 0.1 = 1.72 × 10^{-3} mol

iii Therefore, in 250 cm³ the learner has 1.72×10^{-2} mol or $1.72 \times 10^{-2} \times 84.1$ g = 1.445 g

iv If each learner requires 2.00 g of mixture, then the sodium chloride should contribute 2.00 − 1.45 g = 0.55 g.

v The percentage composition of the mixture = 72.5% $NaHCO_3$ and 27.5% NaCl. If you have 20 learners then you will need 20 × 2.00 g of mixture but allow for more because learners are still in the early stages of learning in detail about quantitative investigations and their technique may not yet be up to scratch.

vi Whatever is decided, there could be differences in the results obtained because the solid mixture may not be homogeneous. The only way to ensure complete homogeneity is to make up a solution of the mixture. This makes a very good discussion point at the end.

vii It is important that learners express the burette readings to ±0.05 cm³. For example, if they get two readings such as 17.00 and 17.10, then the average is 17.05 because burettes usually read to 0.05 cm³, which is one drop.

- Please be aware that learners tend to 'blow out' or expel the last drop of solution from their pipette. The pipette is calibrated so that this last remaining drop is not in the 25.00 cm³ used.

- The end-point of the methyl orange is in fact an orange colour. If a red colour is obtained, then they have overshot.

- The biggest problem is how well you have mixed the sodium hydrogen carbonate and sodium chloride. It is not that big a problem because the apparatus used is overall very accurate and therefore the systematic errors are small. It is a random source of error and a source of an 'open question' at the end of the practical. A systematic error could be the learner who does the same thing wrong for every titration.

🔧 As already mentioned, making the whole mixture into a solution would remove the possibility of random distribution of the solids. Ask them to put forward one way to overcome the problem and see if they come up with a plausible method.

Common learner misconceptions

- The most common error is that learners forget that 25 cm³ is only $\frac{1}{10}$ th of the total amount of solution they have prepared.

Sample results

	Rough titration/cm³	First accurate titration/cm³	Second accurate titration/cm³	Third accurate titration/cm³
Final burette reading/cm³	18.00	35.20	19.80	37.00
Starting burette reading/cm³	0.00	18.00	2.20	19.80
Titre/cm³	18.00	17.20	17.60	17.20

Table 1.4

Answers to workbook questions (using the sample results)

a Change in enthalpy of hydration of copper(II) sulfate

b **i** Volume of 0.100 mol dm⁻³ hydrochloric acid needed to react completely with the sodium hydrogen carbonate present in 25 cm³ of the mixture = 17.20 cm³

ii Number of moles of hydrochloric acid reacting = number of moles of sodium hydrogen carbonate present in 25.00 cm³ = $17.20 \times 10^{-3} \times 0.100 = 1.72 \times 10^{-3}$ = number of moles of sodium hydrogen carbonate present in 25.00 cm³ of solution.

Therefore, in 250 cm³ of solution the number of moles of sodium hydrogen carbonate present = $1.72 \times 10^{-3} \times 10 = 1.72 \times 10^{-2}$ mol

iii Mass of sodium hydrogen carbonate present ($m = n \times M_r$) = $1.72 \times 10^{-2} \times 84.1 = 1.45$ g

iv Total mass of mixture = 2.00 g

v Mass of sodium chloride present in mixture = 2.00 − 1.45 = 0.55 g

vi Percentage of sodium hydrogen carbonate present mixture = $\frac{1.45}{2.00} \times 100\% = 72.5\%$

> **vii** What is the actual percentage composition of the mixture? = 72.5% $NaHCO_3$ and 27.5% $NaCl$
>
> If you have 20 students then you will need 20 × 2.00 g of mixture but allow for more because students are still in the early stages of learning in detail about quantitative investigations and their technique may not yet be up to scratch.
>
> **c** Percentage error =
> $$\frac{|\text{Actual value} - \text{experimental value}|}{\text{Actual value}} \times 100$$
>
> **d** The systematic errors:
>
>   **i** The top pan balance: if 2 readings are taken and the balance reads to 0.01 g then the percentage error for a mass of 2.00 g the percentage error = $2 \times \frac{0.005}{2.00} \times 100\% = 0.5\%$
>
>   **ii** The pipette: if this reads to ± 0.05 cm^3 then the percentage error = $\frac{0.05}{25.00} \times 100\% = 0.200\%$
>
>   **iii** The burette readings
>
> It is important that the students express the burette readings to ± 0.05 cm^3. For example, if they get two readings such as 17.00 and 17.10 then the average is 17.05 because burettes usually read to 0.05 cm^3, which is approximately one drop of solution.
>
> The uncertainty for a burette is ± 0.05 cm^3 for each reading. Therefore, the uncertainty associated with the difference between two burette readings (a titre)
>
> = 2 × 0.05 = ± 0.10 cm^3
>
> Therefore, the error = $\frac{0.10}{17.20} \times 100\% = 0.58\%$
>
> **e** The main random error depends on the homogeneity of the mixture. Another possible error is in the purity of the sodium hydrogen carbonate. Over time it can decompose to give sodium carbonate.
>
> **f** The main contribution to any percentage error is due to the solid mixture not being homogeneous.
>
> **g** The only way to ensure complete homogeneity is to make up a solution of the mixture. This makes a very good discussion point at the end.

Practical investigation 1.4:
Relative atomic mass of calcium by two different methods: molar volume and titration

Introduction

In this investigation, learners will react calcium with water to give hydrogen. The volume of hydrogen given by a known mass of calcium is measured and, using molar ratios, the number of moles of calcium is found and from this the relative atomic mass. The reaction of calcium with water also gives the alkali calcium hydroxide, which is titrated against standard hydrochloric acid. Again, the number of moles of calcium hydroxide (and therefore calcium) is determined and this will give another value for the relative atomic mass.

Skills focus

The following skill areas are developed and practised (see the skills grids at the front of this guide for codes):

MMO	Collection of data and observations: (a), (b), (c), (d) and (e)
	Decisions relating to measurements of observations: (a), (b) (c) and (d)
PDO	Recording data and observations: (c) and (e)
	Display of calculations and reasoning: (a) and (b)
ACE	Data interpretation and sources of error: (g) (h), (i) and (j)
	Drawing conclusions: (c) and (d)
	Suggesting improvements: (a) and (c)

Duration

This investigation is a summative exercise as it uses techniques from Investigations 1.2 and 1.3 and requires the learners to use several formulae and relationships.

Preparing for the investigation

- Of the two sets of apparatus suggested for collecting the gas, the easiest to set up is the one using the gas syringe. However, if gas syringes are not available, then the displacement of water in a measuring cylinder works very well.

Equipment

Each learner or group will need:

- apparatus for measuring gas volumes as used in Investigation 1.2

- small filter funnel for burette
- 50 cm³ burette
- weighing boat
- 150 cm³ conical flask
- wash bottle of distilled water
- burette stand
- 25 cm³ pipette
- white tile
- 250 cm³ beaker
- 25 cm³ measuring cylinder (for water)
- methyl orange indicator in dropper bottle

Access to:
- top-pan balance reading to **at least** two decimal places. A top-pan balance reading to three decimal places is preferable.
- 0.200 mol dm^{-3} hydrochloric acid
- access to **fresh** calcium granules
- distilled water

Safety considerations

- Learners must wear eye protection and tie their hair back if it is long.
- Calcium reacts vigorously with water. Emphasise that learners should not handle it with wet hands.
- Hydrogen is a flammable gas.
- 0.2 mol dm^{-3} hydrochloric acid is an irritant.
- It is important that if learners are using gas syringes they do not clamp the syringe too tightly. Firstly, they could crack the glass and it may also hinder the movement of the piston.
- The calcium hydroxide is an alkali and should be regarded as being corrosive. It should be washed off immediately if spilt on the skin.
- Methyl orange indicator is poisonous. If any is splashed onto skin it should be washed off immediately.

Carrying out the investigation

- One problem that needs to be overcome first is making sure that the measuring cylinder is full of water when it is put in the trough and that none or very little escapes. This can be done by either learners holding their hands over the end of the measuring cylinder or placing a piece of plastic wrap over the open end and then turn the measuring cylinder upside down when it is in the trough. Remember to remove the film before starting the actual measurement.
- The main problem with the practical is the freshness of the calcium. If it is visibly oxidised, then the results will be inaccurate and this is one of the random errors encountered. If the top portion of your calcium looks to be oxidised then use the lower portions. An alternative is that if you know you are going to use calcium for Group II experiments, then as soon as it is bought, divide it up into smaller portions and store in small containers until ready to use. It is the constant exposure to air that leads in the end to its oxidation.
- Make sure that learners have at least two sets of results to analyse. They may struggle on the first set but will get better the more practice they have.
- Once learners have started, then one of the group can do the determination of gas volumes while the other can do the titration. After they have done this once they can swap over.
- Before the practical, a short demonstration with an approximate mass of calcium will give learners some idea of the volumes of gas that they will be dealing with.
- Also, unless there is time for a trial run, learners could be given an idea of the volume of acid required for the titration.
- Evaluation of a practical method always presents problems to learners and they will need help when estimating the percentage error due to using different apparatus.
- If you want to extend the more able learners, you can state that they know what the answer should be and they can work back to see what readings they should get. However, in this case it should be emphasised that the methods are not perfect and therefore cheating will give them fewer marks.

Common learner misconceptions

- Learners may need to be reminded that the calcium hydroxide is formed from the same mass of calcium as in the first method. This fact sometimes becomes lost when learners are doing their calculations.

Chapter 1: Masses, moles and atoms

Sample results

Part 1: Determination by molar volume

Example measurements shown in Table 1.5.

Learner	Mass of Ca/g	Volume of H_2/cm³	At mass of Ca/g mol⁻¹	Burette readings		Vol of HCl/cm³	At mass of Ca/g mol⁻¹
1	0.050	30.00	40.00	2nd	12.50	12.50	40.00
				1st	0.00		
	0.048	28.0	41.10	2nd	23.90	11.40	42.30
				1st	12.50		
2	0.040	24.0	40.0	2nd	10.00	10.00	40.00
				1st	0.00		
	0.055	31.0	42.5	2nd	24.50	14.50	37.70
				1st	10.00		
3	0.060	37.0	38.9	2nd	17.20	15.20	39.50
				1st	2.00		
	0.071	42.0	40.57	2nd	36.20	18.20	39.01
				1st	18.00		

Table 1.5

Answers to the workbook questions (using the sample results)

Weight of calcium = 0.048 g

a i Number of moles of hydrogen formed in first experiment:

$$n_{hydrogen} = \frac{28}{24\,000} = 1.17 \times 10^{-3}\,\text{mol}$$

ii Number of moles of calcium:

$$n_{calcium} = n_{hydrogen} = 1.17 \times 10^{-3}\,\text{mol}$$

iii Relative atomic mass of calcium:

$$A_r(Ca) = \frac{0.048}{1.17 \times 10^{-3}} = 41.0\,\text{g mol}^{-1}$$

b The percentage error in your result

Percentage error =

$$\frac{|\text{Actual value} - \text{experimental value}|}{\text{Actual value}} \times 100$$

The relative atomic mass result for this experiment is 41.0 which should be 40.1.

This gives a percentage error of $\frac{41.0 - 40.1}{40.1} \times 100\%$ = 2.24%

c Systematic errors in the apparatus:

  i The weighing out of the calcium: If you use a top-pan balance reading to ± 0.001 g then the possible error is ½ × 0.001 = 0.0005.

A mass of 0.048 g has a possible error of $\frac{0.0005}{0.048} \times 100\%$ = 1.04%. This will rise to 10.4% if you use a top-pan balance that measures to two decimal places.

  ii The measurement of gas volume

A 100 cm³ measuring cylinder reads to ±2.00 cm³ and therefore has a maximum error of ±1.00 cm³. A volume of 28.0 cm³ has a possible error of $\frac{1}{28.0} \times 100\%$ = 3.60%.

  iii Random errors in the method:

The calcium is possibly oxidised. In this case, the volume of hydrogen will be less than ideal and the value of n will be lower than expected. Therefore $\frac{m}{n}$ will give a value of the relative atomic mass higher than the published value. This method also assumes that the hydrogen is collected at R.T.P.

  iv Improvements to Method 1:

If the calcium is oxidised than some of the mass weighed out is not calcium. The best way round this is to not use the calcium at the top of the container but use the calcium below it because it is less exposed to air.

Part 2: Determination by titration

Answers to the workbook questions (using 2nd set of sample results for learner 1 in Table 1.5)

Weight of calcium = 0.048 g

d The number of moles of hydrochloric acid reacting with the calcium hydroxide:

$n_{calcium} = ½ \, n_{HCl} = ½ \times 11.40 \times 10^{-3} \times 0.200 = 1.14 \times 10^{-3}$

  i The number of moles of calcium hydroxide and therefore the number of moles of calcium:

  $n_{calcium} = ½ \, n_{HCl} = ½ \times 11.40 \times 10^{-3} \times 0.200 = 1.14 \times 10^{-3}$

  ii The relative atomic mass of calcium:

  $A_r(Ca) = \dfrac{0.048}{1.14 \times 10^{-3}} = 42.1 \, g \, mol^{-1}$

e The percentage errors:

  i Weighing out of the calcium: $2 \times (\dfrac{0.0005}{0.048}) \times 100\%$ = 2.08%.

  ii The titrations: The burettes read to $\pm 0.05 \, cm^3$ and therefore in a titration where two readings are made, the error = $2 \times 0.05 = \pm 0.10 \, cm^3$. This means that in the set of readings used, the error

  $= \dfrac{0.10}{11.40} \times 100\% = 0.88\%$

  iii Total systematic error due to apparatus
  = 2.08 + 0.88 = 2.96%

  iv If the value for the relative atomic mass is greater than it should be then $\dfrac{m}{n}$ is greater than it should be and we have overestimated the number of moles of calcium. This may be due to oxidation of the calcium so that n is smaller leading to a greater than expected value for A_r.

f Improvements to Method 2: If the concentration of the hydrochloric acid is reduced then more of it will be required in the titrations. This will reduce the percentage error in the results due to the titrations.

Structure and bonding

Chapter outline

This chapter refers to Chapters 4: Chemical bonding and Chapter 5: States of matter in the coursebook.

In this chapter learners will complete practical investigations on:

- 2.1 Physical properties of three different types of chemical structure
- 2.2 Effect of temperature on the volume of a fixed mass of gas
- 2.3 Effect of pressure on the volume of a fixed mass of gas

Preparing for the investigations

- Learners should be familiar with types of structure and how its structure affects the properties of a substance.
- Learners should be familiar with the behaviour of different types of structure when testing for electrical conductivity and melting point.
- The theory underlying the Gas Laws is fairly straightforward but the practical work will underpin what is learned in the theory lessons.
- If there are enough data loggers for a class practical then learners should be made familiar with how to use them, especially the single-step procedure which is required in this practical. If the Boyle's Law investigation is demonstrated then linking the data logger to a whiteboard would help learners follow the practical.

Practical investigation 2.1:
Physical properties of three different types of chemical structure

Introduction
In this investigation, learners will carry out tests on three substances and make sense of their observations by using their knowledge of structures.

Skills focus
The following skill areas are developed and practised (see the skill grids at the front of this guide for codes):

MMO   Collection of data and observations (a, b, c, d and e)
           Decisions relating to measurements of observations (c) and (d)

PDO   Recording data and observations (a), (c) and (e)

ACE   Data interpretation and sources of error (b)
           Drawing conclusions (c) and (d)

Duration

- This investigation requires ½ hour for the practical work and the rest of the hour for discussion of results and a plenary.

Preparing for the investigation

- The silicon dioxide often contains impurities therefore it needs to be washed thoroughly with distilled water and dried in an oven before use.

Equipment

Each learner or group will need:

- Bunsen burner, tripod, gauze and heatproof mat
- 12 dry test tubes and a test-tube rack
- eight stoppers to fit test tubes
- two graphite rods in a holder
- three spatulas
- three leads and two crocodile clips
- 12 V bulb
- power pack
- wash bottle filled with distilled water
- small evaporating basin
- tongs

Cambridge International AS & A Level Chemistry

Access to:
- cyclohexane
- wax
- white sand
- potassium iodide

Safety considerations

- The cyclohexane is flammable and harmful. It should not be disposed of by flushing down the sink. The best way to remove it is to place the reaction liquids in a large bottle which can then be disposed of safely, or it can be distilled and re-used.
- Cyclohexane must be kept away from naked flames.
- Learners should be told to obey safety instructions especially when heating some of the solids very strongly.

Carrying out the investigation

- The methods used are very simple but they need to be aware that apart from the investigation of electrical conductivity, they need to use small amounts of the three solids.

  Ask learners to research the electrolysis of molten potassium iodide on the internet.

Common learner misconceptions

- Learners need to be aware of what constitutes a **soluble** solid. As stated above they need to use small amounts of solid.

Sample results

Please refer to Table 2.1

Substance	Type of structure	Summary of observations
Wax	Simple molecular	Melts easily therefore low melting point. Does not conduct electricity as a solid or in cyclohexane. Soluble in non-polar cyclohexane but not in water.
Silicon dioxide	Giant covalent	Does not melt therefore very high melting point. Does not conduct electricity as a solid or in water. Does not dissolve in water or cyclohexane.
Potassium iodide	Giant ionic	Melts if heated very strongly; high melting point. Does not conduct in solid state but does so in water. Soluble in water, insoluble in cyclohexane.

Table 2.1

Answers to the workbook questions (using the sample results)

Answers to the workbook questions

a Explain your observations for each of the **three** substances

i Wax

Wax has a simple molecular structure. It does not conduct electricity in a solid or in solution because there are no charge-carrying particles present. It is non-polar and therefore will dissolve in non-polar cyclohexane but not in polar water.

ii Potassium iodide

potassium iodide has a giant ionic structure. It does not conduct electricity as a solid because the ions cannot move and carry the current, but they can do when they are dissolved in water and therefore potassium iodide solution is a conductor. High melting point because of the strong electrostatic attraction between the oppositely charged ions. It will dissolve in polar water but not in non-polar cyclohexane.

iii Silicon dioxide

Silicon dioxide has a giant covalent structure. It has a high melting point because all the strong covalent bonds have to be broken when it melts. Because all the bonds in the giant structure are covalent, it will not dissolve in polar water or in nonpolar cyclohexane.

Practical investigation 2.2:
Effect of temperature on the volume of a fixed mass of gas

Introduction

In this investigation, learners will investigate how changing the temperature affects the volume of a gas. The gas used is air which is a mixture of gases but it is assumed that it will obey the ideal gas laws under these conditions. This is not a qualitative investigation; it is designed to show that extrapolation back to zero volume will give a temperature close to −273 °C.

Skills focus

The following skill areas are developed and practised (see the skill grids at the front of this guide for codes):

MMO	Collection of data and observations: (a, b, c, d and e)
	Decisions relating to measurements of observations: (a, b, c and d)
PDO	Recording data and observations: (a), (c), (d) and (e)
	Display of calculations and reasoning: (a) and (b)
	Data layout: (b), (c), (d), (e) and (f)
ACE	Data interpretation and sources of error: (a), (c) and (e)
	Drawing conclusions: (c) and (d)

Duration

The practical aspects of this investigation require 1 h to complete. The follow-up lesson can be partly allocated to Practical investigation 2.3 because this will take a very short time. A quarter of an hour is easily enough time if learners are familiar with using the data logger.

Preparing for the investigation

- Learners need to know the volume of the flask and the tubing attaching it to the gas syringe. If the round bottom flasks are identical for all groups then one determination of the volume before the practical lesson will suffice. The method used is described in the practical method. The permanent marker pen can be used to show how far the stopper to the flask protrudes into the neck.

- The volume of the tubing can be found by filling it with water and measuring the water required to fill it. Another way of doing it is to measure the internal diameter of the tubing and its length and using the following formula:

    Volume = (πr^2 × length of tube) where r is the internal radius of the tube.

- If the same length of tubing is used for each group then this value can be given to the whole class.

Equipment

Each learner or group will need:

- Bunsen burner, tripod, gauze and heatproof mat
- 100 cm^3 round-bottomed flask
- stopper for flask attached to a short length of plastic or rubber tubing
- 100 cm^3 measuring cylinder
- permanent marker pen
- dropper
- 100 cm^3 gas syringe or apparatus to measure gas volume by displacement of water (see Skills chapter)
- metal container for heating water
- thermometer reading to 110 °C
- either a stirring rod or a small 'paddle' for stirring water in metal container
- water supply

Safety considerations

- Towards the end of the experiment there will be hot water which needs to be stirred on a tripod and gauze. Therefore, learners need to be very careful when they stir.

Carrying out the investigation

- It will take a little while for learners to understand how much heating is required to give a small temperature rise in the water and inevitably some of them will go past their desired temperature. Firstly, make sure they realise that it is not absolutely essential for them to get a temperature such as 30 °C or 40 °C. A temperature difference of 1 or 2° is not a disaster. If, however, they do go way past a desired temperature, then as described in the method they can add a small amount of cold water to get back near to the desired temperature.

- When analysing their results, learners may find that there is some flexibility when drawing their line. One way around this is to draw small circles round their points and sometimes this helps with ascertaining the best line to draw. The other way is to export their results into an Excel document and then draw a scatter chart to give them the desired line.

- The instructions in the method do help learners draw their line by stating the scales that need to be adhered to when drawing their graph.

    Ask learners to write their own law using their results.

Common learner misconceptions

- Learners often still think in terms of °C when talking about temperature rather than in Kelvin. Therefore, learners often wonder why their line does not go through the origin.

Sample results

Please refer to Table 2.2

Temp/°C	18	32	39	45	54	60	65	72	80	85
Reading on syringe/ml	0	4	7	9	12	14	16	18	21	22
Total volume of gas/cm³	100	104	107	109	112	114	116	118	121	122

Table 2.2

Answers to the workbook questions (using the sample results)

a Figure 2.1 shows the sample results plotted.

   i Check the best-fit line is correct

   ii Temperature where volume is zero = [–290 °C]

b The extrapolated value for the temperature when the volume of the gas is zero is approximately –290 °C. This gives an experimental error of 6.22%.

c The main sources of error in the experiment are the stirring and the synchronisation between the temperature and the volume measurement.

d The name given to the temperature when the volume is zero is absolute zero.

e A reasonable law is that the volume of gas is directly proportional to the temperature of the gas if a scale reading from –290 °C is used for the temperature scale.

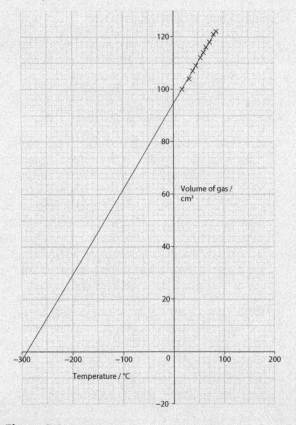

Figure 2.1

Practical investigation 2.3:
Effect of pressure on the volume of a fixed mass of gas

Introduction
In this investigation, learners or the teacher are asked to use a data logger. It is likely that there will be insufficient data loggers for a class practical and therefore it can be viewed as a demonstration. The practical requires the manipulation of data in order to establish a relationship between pressure and the volume of a fixed mass of gas. The gas used is air.

Skills focus
The following skill areas are developed and practised (see the skill grids at the front of this guide for codes):

MMO	Collection of data and observations: (b) and (c)
PDO	Recording data and observations: (a), (c) and (e)
	Display of calculations and reasoning: (a) and (b)
	Data layout: (b), (c), (d) (e) and (f)
ACE	Data interpretation and sources of error: (a)
	Drawing conclusions: (a) and (c)

Duration
- Although Practical 2.2 requires one hour for the practical work to be completed, the follow-up lesson can be partly allocated to this investigation because this will take a very short time. Fifteen minutes is easily enough if learners are familiar with using the data logger.

Equipment
Each learner or group will need:
- a laptop or other device that will interface with a data logger and run the software required
- a pressure data logger with any software required
- a 60 cm^3 plastic syringe attached to a small length of plastic tubing which will fit the inlet to the pressure data logger.

Safety considerations
- The only thing that can possibly happen is that the tube might come off during the experiment but this is not likely to cause any great problems in terms of safety.

Carrying out the investigation
- When using the ideal gas equation, learners have to realise that the unit of volume is m^3 and not cm^3. The units on the data logger are probably given in kPa and they should also realise that this needs converting to Pa.

  The more able learners will realise that $P \alpha \frac{1}{V}$ and that PV = constant. Therefore, once the results are obtained they can be asked to plot the results in any way they wish. The correct way is to plot $\frac{1}{V}$ (horizontal axis) against P (vertical axis). The more able learners could also be asked to use an Excel spreadsheet to process the results.

Sample results
A sample set of results is shown in Table 2.3:

Volume of gas/cm^3	$\frac{1}{volume}$	Pressure of gas /kPa
60	0.0167	102.1
55	0.0182	108.6
50	0.0200	117.4
45	0.0222	127.2
40	0.0250	141.9
35	0.0286	155.2
30	0.0333	176.2
27	0.0370	190

Table 2.3

Answers to the workbook questions (using the sample results)

a $P \alpha \dfrac{1}{V}$

b Refer to data in Table 2.3.

c Please refer to Figure 2.2, obtained from one set of results.

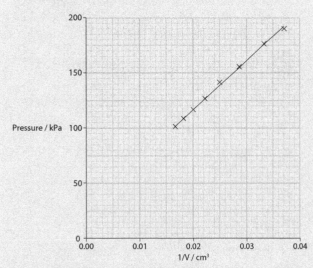

Figure 2.2

Note the R^2 value of 0.9983 is very close to 1 and therefore there is a great deal of certainty that it is a straight line.

d The results give a straight line and therefore the slope of the line is constant.

e The slope of the line $= \dfrac{P}{1/V} = PV = $ constant

f If we take the first values from this set of results

$P = 102.1 \text{ kPa} = 1.021 \times 10^5 \text{ Pa}$

$V = 60 \text{ cm}^3 = 60 \times 10^{-6} \text{ m}^3$; $n = \dfrac{60}{24\,000} = 0.0025 \text{ mol}$;

g The experiment was conducted at 20 °C or 293 K

$R = \dfrac{PV}{nT} = \dfrac{1.021 \times 10^5 \times 60 \times 10^{-6}}{0.0025 \times 293} = 8.36 \text{ J mol}^{-1} \text{K}^{-1}$

h The value of R given in the literature is 8.31 J mol^{-1} K^{-1}

The percentage error $= \dfrac{8.36 - 8.31}{8.31} \times 100\% = 0.60\%$

i If we substitute these units into the equation for the units of R we have:

$R = \dfrac{PV}{nT}$; 1 Pa = 1 N m^{-2} (1 Newton per m^2);

Units of R = N m^{-2} × m^3/mol × K = N m mol^{-1} K^{-1}

1 N m = 1 J

Therefore, units of R = J mol^{-1} K^{-1}

Chapter 3:
Enthalpy changes

Chapter outline

This chapter relates to Chapter 6: Enthalpy changes in the coursebook

In this chapter, learners will complete practical investigations on:

- 3.1 Enthalpy change for the reaction between zinc and aqueous copper(II) sulfate solution
- 3.2 Enthalpy change of combustion of alcohols
- 3.3 Enthalpy change of a thermal decomposition
- 3.4 Change in enthalpy of hydration of copper (II) sulfate

Preparing for the investigations

- Learners should be familiar with terms such as enthalpy and the various definitions associated with this topic. They need to understand these definitions and how they influence the handling of data. For example, standard enthalpy changes are always expressed in terms of kJ mol^{-1}.
- Learners can refer to the Skills chapter for the theory behind the temperature–time graph required for Practical Investigation 3.1.
- Learners should be confident in their conversion of J to kJ and realise that their initial heat calculations are expressed in J.
- Hess' Law investigations are straightforward to carry out but learners often do not understand the underlying theory behind what they are doing.

Practical investigation 3.1:
Enthalpy change for the reaction between zinc and aqueous copper(II) sulfate

Introduction

This investigation requires learners to draw temperature–time graphs and use these to determine the temperature changes in the reaction. The two investigations are concerned with the same reaction but use different limiting reactants for each determination.

The reaction taking place is:

$Zn(s) + CuSO_4(aq) \longrightarrow ZnSO_4(aq) + Cu(s)$

Or more accurately $Zn(s) + Cu^{2+}(aq) \longrightarrow Zn^{2+}(aq) + Cu(s)$

Skills focus

The following skill areas are developed and practised (refer the skills grids at the front of this guide for codes):

MMO	Collection of data and observations (a), (b), (c) (d) and (e)
PDO	Recording data and observations (a) and (e) Display of calculations and reasoning (a) and (b) Data layout (a) (b), (c), (d), (e) and (f)
ACE	Data interpretation and sources of error (a) and (c) Drawing conclusions (c)

Duration

- The practical work will take approximately half an hour to complete.
- The necessary introductions and plenaries will also take half an hour in total so one hour will probably be sufficient.

Preparing for the investigation

- This is probably the first time learners will have gathered data for a temperature–time graph and they may not get it right straightaway.

Equipment

Each learner or group will need:

- two small polystyrene beakers
- glass beaker large enough to hold the polystyrene beakers
- −10 to 110 °C thermometer
- 25 cm^3 measuring cylinder

- plastic covers for polystyrene beakers
- a small spatula
- two weighing boats
- 1 mol dm^{-3} copper(II) sulfate solution
- zinc powder
- a top-pan balance that reads to at least two decimal places

Safety considerations

- The copper(II) sulfate is both an irritant and harmful and the zinc is flammable but there is no reason for using Bunsen burners.
- Be careful when disposing of any copper(II) sulfate because it is regarded as an environmental hazard.

Carrying out the investigation

- The reaction is quite a vigorous one and the lid of the polystyrene beaker does need to be stable and intact when the mixture is swirled.
- The second part of the investigation has the copper(II) sulfate as the limiting reactant and this necessitates the copper(II) sulfate being pure and the concentration of the solution being accurate.

  Initially, some learners do not understand the reasons for doing the experiment using a temperature–time graph; after they have actually done the experiment, however, and the results are explained they can usually understand the rationale.

  Some learners may need help with the calculation but it is probably more important that they complete both practical assignments before they start the calculations.

  The more able learners might find some value in working out the actual enthalpy change for the reaction.

  The reaction may be summarised by the ionic equation:

  $Cu^{2+}(aq) + Zn(s) \longrightarrow Cu(s) + Zn^{2+}(aq)$

  The accepted value for $\Delta H_{reaction} = -219\,kJ\,mol^{-1}$

The more able learners can calculate their values for the enthalpy change of reaction and work out their percentage accuracy using the theoretical results above and their experimental results.

Common learner misconceptions

- Learners must be able to convert heat/enthalpy changes from J to kJ. Occasionally they fail to do so.

Sample results

Part 1

Table 3.1 gives an idea of the results learners should end the investigation with.

Copper(II) sulfate was in excess. The mass of the zinc was:

$0.66\,g = \dfrac{0.66}{65.4} = 0.010\,mol$

The number of moles of copper(II) sulfate = C x V = 1.00 x 0.025 = 0.025 mol.

Therefore, the copper(II) sulfate is in excess and the limiting reactant is the zinc. This means that the number of moles reacting = 0.010 mol

Time /min	Temp. /°C
0	19
1	19
2	19
3	X
4	34.5
5	36
6	36
7	35
8	34.5
9	34
10	33.5

Table 3.1

Chapter 3: Enthalpy changes

Answers to the workbook questions (using the sample results)

Part 1

a See Figure 3.1

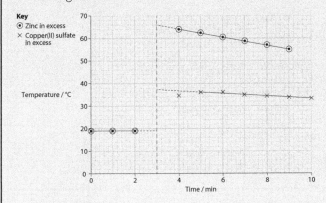

Figure 3.1

When the graph is plotted the initial temperature is 19 °C and the maximum temperature measured by extrapolation is 37.6. The thermometer only reads to 1 °C so this temperature is rounded up to 38 °C.

The temperature change = 38 − 19 = 19 °C.

b Enthalpy change: $q = m \times c \times \Delta T = 25 \times 4.18 \times 19 = 1985.5$ J

c The number of moles of $CuSO_4$ present: $n = C \times V = 0.100 \times 0.025 = 0.0250$ mol

d The number of moles of zinc present: $n = \dfrac{m}{A_r} = \dfrac{0.66}{65.4} = 0.0101$

e In this reaction the zinc reacts with an equal number of moles of copper(II) sulfate. In this experiment there are fewer moles of zinc than copper(II) sulfate so the zinc is the limiting reactant.

f The standard enthalpy change in kJ mol⁻¹

0.101 mol of reactants produce −1985.5 J of heat energy

Therefore, 1 mol of reactants produce $\dfrac{1985.5}{0.0101}$

$\Delta H = -197$ kJ mol⁻¹. This is the standard enthalpy change of reaction. The thermometer only reads to two significant figures and therefore this value can be round up to −200 kJ mol⁻¹.

Sample results

Part 2

Table 3.2 shows the results when zinc is in excess (6.50 g = 0.10 mol). This means that the number of moles reacting = 0.025 mol.

Time /min	0	1	2	3	4	5	6	7	8	9
Temp. /°C	19	19	19	X	64	62.5	60.5	59	57	55

Table 3.2

Answers to the workbook questions (using the sample results)

Part 2

a The number of moles of $CuSO_4$ present: $n = C \times V = 0.100 \times 0.025 = 0.0250$ mol

b The number of moles of zinc present: $n = \dfrac{m}{A_r} = \dfrac{6.6}{65.4} = 0.101$

c See graph drawn in answer to Part 1.

d When the graph is drawn for these results, the initial temperature is 19 °C and the maximum temperature is 66 °C. This means that $\Delta T = 66 - 19\,°C = 47\,°C$

e The enthalpy change: $q = m \times c \times \Delta T = 25 \times 4.18 \times 47 = 4911.5$ J

f The number of moles of $CuSO_4$ present: $n = C \times V = 1.00 \times 0.0250$ mol = 0.0250 mol

g The number of moles of zinc present

$n = \dfrac{m}{A_r} = \dfrac{6.50}{65} = 0.0994$

h In this experiment there is a greater number of moles of zinc and therefore the limiting reactant is the copper(II) sulfate solution.

i The standard enthalpy change in kJ mol⁻¹

∴ The standard enthalpy change of reaction
$\Delta H^{\ominus}_r = \dfrac{4911.5}{0.025} = -19\,6460$ J = −196 kJ mol⁻¹ (3 s.f.)

Note: the thermometer reads to just two significant figures so this can be rounded up to −200 kJ mol−1.

> **j** The reaction is the same both in both experiments and the values are expressed in terms of 1 mol.
>
> **k** The accepted value for $\Delta H^\ominus$, is $-219\,\text{kJ mol}^{-1}$. Therefore, if we use $-200\,\text{kJ mol}^{-1}$ the percentage error for our results $= \dfrac{219 - 200}{219} \times 100\% = 8.7\%$
>
> **l** The highest degree of error will be for Part 1 because the lower temperature rise is recorded and the smallest mass of zinc is weighed.
>
> The balance weighs to 0.010 g and therefore its maximum error is ± 0.005 g.
>
> Therefore, the maximum percentage error from weighing $= 2 \times \dfrac{0.005}{0.66} \times 100\% = 1.52\%$
>
> The thermometer can be read to 0.5 °C, therefore the temperature change = 19 ± 1 °C.
>
> Therefore, the maximum percentage error
> $= \dfrac{1}{19} \times 100\% = 5.3\%$
>
> The volume can be read to 0.5 cm³
>
> Therefore percentage error from volume measurement $= \dfrac{0.5}{25} \times 100\% = 2\%$
>
> The total percentage error from apparatus measurement = 1.52 + 5.30 + 2.00% = 8.82%
>
> **m** The main non-systematic error is obviously the heat loss through the apparatus.

Practical investigation 3.2:
Enthalpy change of combustion of alcohols

Introduction

This practical is as much about what is wrong with it as what is good about it. Ideally, a bomb calorimeter would be used to find the enthalpy of combustion of alcohols. In this practical, however, spirit burners or micro burners are used.

For all four alcohols the temperature of water is raised by the same amount each time. This means that the heat /enthalpy change will be the same each time because the apparatus being used is identical. The apparatus set-up is a very simple one. Heat-resistant pads can be used to make the apparatus draught-free by placing them around the spirit burner and calorimeter. If glass calorimeters are used then the specific heat capacity should be changed but the calculation remains the same.

Skills focus

The following skill areas are developed and practised (refer the skills grids at the front of this guide for codes):

MMO	Collection of data and observations (a), (b), (c). (d) and (e)
	Decisions relating to measurements of observations (c)
PDO	Recording data and observations (a)
	Display of calculations and reasoning (a) and (b)
	Data layout (b), (c), (d), (e) and (f)
ACE	Data interpretation and sources of error (g) (h), (i) and (j)
	Drawing conclusions(c)

Duration

- The practical can be completed easily within one hour.
- Spirit burners can be allocated to specific alcohols and can be shared amongst the groups of learners. Groups of two are ideal.

Preparing for the investigation

- Learners need to know the definition of enthalpy of combustion

Equipment

Each learner or group will need:

- spirit burners containing the four alcohols
- copper wire stirrer
- clamp stand, boss and clamp
- at least two heat-resistant pads
- thermometer
- 100 cm³ measuring cylinder
- lid/cover for spirit burner
- wooden splint

Access to:

- a top-pan balance reading to at least two decimal places – two balances strategically placed would be ideal
- a supply of water
- a Bunsen burner (for lighting splints)

Safety considerations

- They must wear eye protection at all times.
- All the alcohols are flammable.
- All the alcohols should be treated as harmful.
- When weighing the alcohols learners must transport them to the balance on a heat-resistant pad.

Carrying out the investigation

- Learners should realise that the heat required to heat up the calorimeter must be taken into consideration.
- Learners need to realise that as the spirit burners will almost certainly be shared amongst the group; they could well be holding up other groups if they do not complete their determination in a reasonable time.
- The biggest problem is getting the flame to the same height for each burner and adjusting the calorimeter position so that its bottom is the same distance from the flame each time.
- The temperature rise advocated in the workbook is 20 °C, which can be reached in a very short time – about 1 min. Extensive and vigorous stirring is required in each determination.
- Learners have had about four or five investigations by now in which they have been shown how to calculate the errors due to their equipment. In this practical, they should at least be able to work out the percentage error of their results compared with the values available in scientific literature. After they have calculated the errors due to their apparatus, there should still be a certain percentage error which is not accounted for.
- The most obvious source of error is heat loss through the apparatus. Also, the combustions of the alcohols, especially those with a higher relative molecular mass, are not complete and the learner should be reminded of the definition of standard enthalpy of combustion.
- This incomplete combustion can be demonstrated by inspection of the underside of the calorimeter, where a carbon deposit is a good sign of incomplete combustion.
- One way to extend the learner is to encourage them to use spreadsheets in order to automate their calculations and save themselves time. One could say that this is taking them away from carrying out the calculations. However, in constructing their spreadsheet and getting it working they are in fact doing the calculations and putting in formulae that they need to understand.
- To extend the more able learners, they can either take their values for the standard enthalpies of combustion or the values available in the literature and plot them against the relative molecular mass. From their graph they can find the enthalpy change when $-CH_2-$ is burned and from this calculate the bond enthalpy of the C–H bond.

Common learner misconceptions

- Learners sometimes fail to remember that they are heating up the calorimeter as well as the water. Because of this they must remember that the calorimeter has a different specific heat capacity to the water. The specific heat capacity of copper is $0.385 \, J \, g^{-1} \, K^{-1}$ and that of glass is $0.84 \, J \, g^{-1} \, K^{-1}$.

Answers to the workbook questions (using the sample results)

a The standard enthalpy changes of combustion for all four alcohols

Mass of copper calorimeter = 198.00 g

Enthalpy change for heating up 100 g of water by 20 °C in this copper calorimeter

$= 0.385 \times 20 \times 198.00 + 4.18 \times 20 \times 100$

$= 1524.6 + 8360 = 9884.6 \, J$

This enthalpy change is the same for all four alcohols. In the results given below spirit burners were not available and microburners were used. The time taken for each determination was approximately 1–2 mins

Table 3.3 shows the results for all four alcohols taken from a spreadsheet.

Alcohol	Mass of burner + alcohol before burning	Mass of burner + alcohol after burning	Mass burned	RMM of alcohol	no. of moles burned	Enthalpy change/J	Standard enthalpy change in kJ mol^{-1}
methanol	5.41	4.79	0.62	32	0.019375	9884.6	-510.2
ethanol	6.05	5.57	0.48	46	0.010435	9884.6	-947.3
propan-1-ol	6.20	5.79	0.41	60	0.006833	9884.6	-1446.5
butan-1-ol	6.27	5.87	0.40	74	0.005405	9884.6	-1828.7

Table 3.3

b The percentage errors for each alcohol are shown in Table 3.4.

Alcohol	Standard enthalpy change in kJ mol^{-1}	Literature values for standard enthalpy of combustion in kJ mol^{-1}	percentage error
methanol	-510	-726	29.7
ethanol	-947	-1367	30.7
propan-1-ol	-1447	-2021	28.4
butan-1-ol	-1829	-2676	31.7

Table 3.4

c The maximum percentage error from apparatus shown in Table 3.5

Apparatus/reading	Reading error	Reading taken	Percentage error	Comments
Top-pan balance reads to 0.01 g error = 0.005	0.01	0.62	1.61	There are two mass readings with an error of ± 0.005 g each time
Measuring cylinder reading to ± 2 cm^3	1.00	100	1.00	The measuring cylinder measures to 2 cm^3 therefore uncertainty is ±1 cm^3
Thermometer reading to ± 0.5	1.00	20.00	5.00	Two thermometer readings taken – both giving maximum error of ± 0.5 °C – therefore total error is 1.0 °C
		Total %	7.61	

Table 3.5

d Measurement of uncertainty for the mass of alcohol burned for each alcohol.

Methanol   Percentage error = $\dfrac{0.01}{0.62} \times 100\% = 1.61\%$

Ethanol   Percentage error = $\dfrac{0.01}{0.48} \times 100\% = 2.08\%$

Propan-1-ol   Percentage error = $\dfrac{0.01}{0.41} \times 100\% = 2.44\%$

Butan-1-ol   Percentage error = $\dfrac{0.01}{0.40} \times 100\% = 2.50\%$

Chapter 3: Enthalpy changes

> **e** Maximum percentage error for one alcohol
>
> Example = methanol
>
> Total error due to measuring apparatus = 7.61%
>
> **f** For methanol, the difference between the total percentage error and the error due to the apparatus is 29.7 − 7.61 = 22.09%. This considerable difference is probably due to incomplete combustion of the alcohol and heat loss through conduction (through sides of calorimeter) and convection (hot waste gases not heating up calorimeter).

Practical investigation 3.3:
Enthalpy change of thermal decomposition

Introduction

The enthalpy change for some reactions is impossible to measure. Because thermal decomposition is an endothermic reaction, it is impossible to find the heat change directly. The only way to find these changes is to use Hess' Law. In this investigation we look at the thermal decomposition of potassium hydrogen carbonate.

$$2KHCO_3(s) \longrightarrow K_2CO_3(s) + CO_2(g) + H_2O(l)$$

Skills focus

The following skill areas are developed and practised (see the skill grids at the front of this guide for codes):

MMO	Collection of data and observations (a), (b), (c), (d) and (e)
	Decisions relating to measurements of observations (d)
PDO	Recording data and observations (a) and (c)
	Display of calculations and reasoning (a) and (b)
ACE	Data interpretation and sources of error (a), (c) (d) and (e)
	Drawing conclusions (c) and (d)

Duration

- The practical can be completed easily within half an hour.
- Learners can work individually.
- If a lesson lasts for one hour then the majority of the time can be spent explaining the theory behind the method used and, in the plenary, how they are going to calculate the changes.

Preparing for the investigation

- Learners need to know the theory behind Hess' Law and how it can be used to determine enthalpy changes that otherwise would be impossible to determine.
- Learners need to revise reactions between acids and carbonates or hydrogen carbonates.

Equipment

Each learner or group will need:

- polystyrene beaker and lid with hole for thermometer
- glass beaker to hold the polystyrene beaker
- thermometer – one reading from -10 to 50 °C with 0.2 °C divisions is preferable
- spatula
- weighing boats
- 50 cm^3 measuring cylinder
- cotton wool to act as extra insulation

Access to:

- a top-pan balance reading to at least two decimal places – two balances strategically placed would be ideal.
- a supply of water
- 2.00 mol dm^{-3} hydrochloric acid
- potassium hydrogen carbonate and potassium carbonate

Cambridge International AS & A Level Chemistry

Safety considerations

- Learners must wear eye protection at all times
- The acid is an irritant
- In the reactions there is quite a lot of effervescence and therefore care must be taken in replacing the lid as soon as the solids are added to the acid to minimise exposure to acid spray.

Carrying out the investigation

- Many learners do not use twice the enthalpy change ΔH_1 in their calculations. They should also realise that the sign given to the enthalpy change (plus or minus) is vitally important to the final result.

- A number of learners struggle with Hess' law and therefore practice in these calculations will obviously help them overcome these difficulties

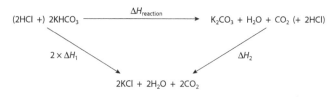

Figure 3.2

Encourage learners to use spreadsheets in order to automate their calculations and save themselves time. If you are worried that this will take them away from carrying out the calculations themselves, remember that in constructing their spreadsheet and getting it to work they are in fact doing the calculations and putting in formulae that they need to understand.

Common learner misconceptions

- The Hess cycle used for this practical investigation is shown in Figure 3.2.

Answers to the workbook questions (using the sample results)

a–f The results shown in Table 3.6 are from a typical laboratory investigation of this topic.

Mass of $KHCO_3$ = 2.55 g   Mass of K_2CO_3 = 3.46 g

	Initial temp. °C	Final temp. °C	Change in temp. °C	Enthalpy change (q)/J	Mass of $KHCO_3$/g	Relative formula mass/ g mol^{-1}	Number of moles	Standard enthalpy change
Reaction 1	18	14.4	−3.6	+752.4	2.55	100.1	0.025475	+29.5 kJ mol^{-1}
Reaction 2	17.9	22.2	+4.3	−898.7	3.46	138.2	0.025036	−35.9 kJ mol^{-1}

Table 3.6

g The standard enthalpy change for the reaction:

Using Hess' Law $\Delta H_r + \Delta H_2 = 2\Delta H_1$;

$\Delta H_r = 2\Delta H_1 - \Delta H_2 = 2 \times 29.5 - (-35.9) = +94.9$ kJ mol^{-1}

The accepted values using the standard enthalpies of formation = +93.4 kJ mol^{-1}

h The percentage error = $[\frac{94.9 - 93.4}{93.4}] \times 100\% = 1.6\%$

i Maximum percentage error

Reaction 1

The thermometer reads to ± 0.2 °C and therefore the uncertainty is ±0.1 °C. There are two temperature readings and therefore the total uncertainty is 0.2 °C

The percentage error = $(\frac{0.2}{3.6}) \times 100\% = 5.56\%$

For the weighing, there are two readings being made and for each one the uncertainty is ±.005 g.

Therefore the percentage error = $2 \times \frac{0.005}{2.55} \times 100\% = 0.39\%$

For measurement of acid using measuring cylinder the measuring cylinder measures to 1 cm³ and therefore uncertainty = ± 0.5 cm³

Percentage error = $(\frac{0.5}{50}) \times 100\% = 1\%$

Therefore, the total percentage error for reaction 1 = 5.56 + 0.39 + 1 = 6.95%

Reaction 2

Percentage error from temperature measurement = $(\frac{0.2}{4.3}) \times 100\% = 4.65\%$

Percentage error from weighing = $2 \times \frac{0.005}{3.46} \times 100\% = 0.29\%$

Percentage error due to measurement of acid = 1%

Therefore, total percentage error = 4.65 + 0.29 + 1.0 = 5.94%

The total possible error due to measuring apparatus = 5.94 + 6.95 = 12.89%

This means that the actual error for the experiment (1.6%) is well within the error due to the measuring apparatus.

Practical investigation 3.4:
Change in enthalpy of hydration of copper (II) sulfate

Introduction

This practical will complete this series of experiments because it allows learners to both use techniques they have already encountered and enables them to make calculations based on Hess's law.

The reaction studied is the hydration of copper(II) sulfate:

$$CuSO_4(s) + 5H_2O(l) \longrightarrow CuSO_4.5H_2O$$

The Hess cycle used is shown in Figure 3.3.

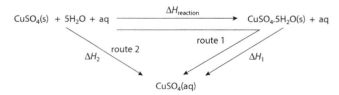

Figure 3.3

Skills focus

The following skill areas are developed and practised (refer the skills grids at the front of this guide for codes):

MMO   Collection of data and observations (a), (b), (c), and (d)

           Decisions relating to measurements of observations (b)

PDO   Recording data and observations (a), (c), (d) and (e)
           Display of calculations and reasoning (a) and (b)
           Data layout (a), (b), (c), (d), (e) and (f)

ACE   Data interpretation and sources of error (a), (c), (d) and (e)
           Drawing conclusions (c) and (d)

Duration

This practical takes one hour to compete.

Preparing for the investigation

- Preparation of anhydrous copper(II) sulfate. To save time and reduce inaccuracies, it is best to take some copper sulfate crystals and heat them in an oven which is set at about 200 °C. It is advisable to check the oven beforehand by heating a small quantity of crystals to check the accuracy of the oven's temperature.

- Each learner or group of learners will need something in excess of 4 g to weigh out for their 0.025 mol of the anhydrous salt. Therefore, this must be taken into account when deciding how much needs to be roasted in the oven. The solid will need to be stirred at various times to make sure that the loss of water is uniform throughout the solid.

Cambridge International AS & A Level Chemistry

Equipment

Each learner or group will need access to:

- two polystyrene beakers plus lids
- thermometer which reads from -10 °C to 50 °C in 0.2 °C divisions
- spatula
- wash bottle containing distilled water
- glass beaker large enough to hold the polystyrene beakers
- cotton wool to improve the insulation of the polystyrene beakers
- a 50 cm³ measuring cylinder
- weighing boat x 2
- Top-pan balance which reads at least to two decimal places
- anhydrous copper(II) sulfate
- hydrated copper(II) sulfate crystals
- distilled water
- paper towels

Safety considerations

- Eye protection must be worn at all times during this experiment.
- The copper(II) sulfate solution is an irritant and copper(II) sulfate is an environmental poison; any solution formed should be poured into a bottle. This copper sulfate solution can be used to crystallise out pure copper sulfate which can be used for other experiments.

Carrying out the investigation

- The temperature change for the dissolving of the hydrated copper(II) sulfate crystals is small and therefore if at all possible a thermometer (or temperature data logger) reading to 0.2 °C should be used.
- Some learners will still have problems with understanding the Hess cycle. The cycle, however, is quite easy to understand. They will need help with the amount of water present in the hydrated crystals.

Learners have had a reasonable amount of experience now using the techniques and theory utilised in this practical. Therefore, it may be a good idea to use this practical investigation to gauge their progress.

Ask learners to explain how the apparatus could be improved (e.g. Dewar flasks could be used). A joulemeter and immersion heater can be used to measure the actual amount of energy required to heat up the flask by the temperature measured in the reaction. This then accounts for the energy needed to heat up the flask as well as the water. Ask them to write an explanation of how that set-up works and how it is an improvement on their own set-up.

Common learner misconceptions

- It is difficult for some learners to understand why it is impossible to determine the enthalpy change for the hydration of anhydrous copper(II) sulfate.

Sample results

The following results shown in Table 3.7 were obtained for this experiment for ΔH_2

Mass of anhydrous copper(II) sulfate = 3.99 g = $\frac{3.99}{159.6}$ = 0.025 mol

Time /min	Temperature/°C
0	17.5
1	17.6
2	17.8
3	18.1
4	X
5	25.4
6	25.8
7	25.7
8	25.5
9	25.3
10	25.1

Table 3.7

Chapter 3: Enthalpy changes

Answers to the workbook questions (using the sample results)

Part 1: For the determination of ΔH_2.

a See Figure 3.4

b Using the graph the initial temperature = 18.2 °C and the final temperature = 26.4 °C

c Temperature change = 8.2 °C

d ∴ Enthalpy change = 50 × 4.18 × (26.2 − 18.2) = −1672 J

e Standard enthalpy change for ΔH_2 = = $(\frac{-1672}{0.025}) \div 1000$ = −66.9 kJ mol^{-1}

For ΔH_1; Mass of copper(II) sulfate crystals = 6.24 g

= $\frac{6.24}{249.6}$ = 0.025 mol

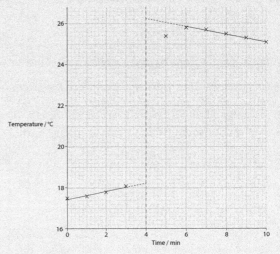

Figure 3.4

Part 2: For the determination of ΔH_1.

a The initial temperature = 18.0 °C and the final temperature = 17.0 °C

b ∴ Enthalpy change = 50 × 4.18 × (18.0 − 17.0) = +209 J

c Standard enthalpy change for ΔH_2 = (+209/0.025) ÷ 1000 = +8.36 kJ mol^{-1}

d ∴ $\Delta H_{reaction} = \Delta H_2 - \Delta H_1$ = −66.9 − (+8.36) = −75.3 (three significant figures)

e The accepted value is −78.2 ∴ percentage error = $\frac{78.2 - 75.3}{78.2}$ × 100% = 3.71%

f The errors due to the apparatus are shown in Table 3.8.

Apparatus/reading	Reading to	Reading taken	Percentage error	Comments
Top-pan balance reads to 0.01 g error = 0.005	0.01	3.99	0.25	There are two mass readings with an error of ± 0.005 g each time
	0.01	6.24	0.16	
Measuring cylinder reading to ± 1 cm^3	1.0	50.0	1.00	The uncertainty is ± 0.5 cm^3
	1.0	50.0	1.00	
Thermometer reading to ± 0.2	0.2	8.2	2.44	Two thermometer readings taken for each experiment – both giving maximum error of ± 0.1 °C therefore total error is 0.2 °C
	0.2	1.0	20.00	
		Total %	24.85	

Table 3.8

The potential errors due to the measuring apparatus can explain the percentage error in the experiment.

Redox reactions

Chapter outline

This chapter refers to Chapter 7: The properties of metals in the coursebook

In this chapter, learners will complete practical investigations on:

- 4.1 Understanding redox (I): investigating a reactivity series and displacement reactions
- 4.2 Understanding redox (II): investigating further reactions

Preparing for the investigations

- Learners need to think of redox reactions as either changes in oxidation number or loss and gain of electrons.
- Learners should be aware of how to test for cations and the results of those tests.

Practical investigation 4.1: Understanding redox (I): investigating a reactivity series and displacement reactions

Skills focus

The following skill areas are developed and practised (refer the skills grids at the front of this guide for codes):

MMO	Collection of data and observations (b), (c) and (e) Decisions relating to measurements of observations (a), (d) (e) and (f)
ACE	Drawing conclusions (c) and (d)

Duration

The practical should take about one hour to complete.

Equipment

Each learner or group will need:

- 10 test tubes
- two test-tube racks
- six droppers
- eye protection
- wooden splint
- Bunsen burner and heatproof mat
- small spatula
- small glass filter funnel and three filter papers

Access to:

- $0.500 \, mol \, dm^{-3}$ copper(II) nitrate solution
- $0.500 \, mol \, dm^{-3}$ zinc nitrate solution
- $2.00 \, mol \, dm^{-3}$ hydrochloric acid solution
- magnesium ribbon
- magnesium powder
- $2.00 \, mol \, dm^{-3}$ sodium hydroxide solution
- iron powder
- zinc powder

Safety considerations

- The copper(II) nitrate solution is both an irritant and harmful.
- The sodium hydroxide is corrosive at this concentration.
- The hydrochloric acid is an irritant.
- The magnesium powder and ribbon and the zinc powder are flammable.
- Learners should wear eye protection at all times.

Carrying out the investigation

- The practical work is very simple but instead of just accepting the identities of the products formed, the learners are asked to confirm their conclusions by testing for cations.

Chapter 4: Redox reactions

- For each of the reactions, it is important to add an excess of the more reactive metal so that the reaction goes to completion. In Reaction 1, if there is still acid present after the reaction then more sodium hydroxide solution is required before a positive result is obtained.

  ⚙ Learners can be asked to make up their own reaction mixture along with the chemical test to identify the ion formed in the reaction.

Common learner misconceptions

- If learners understand the concept of oxidation and reduction in terms of changes in oxidation number or by loss and gain of electrons then there is very little to misunderstand about the practical. The colours of the precipitates formed are not a problem because they are different colours.

Sample results

Reaction	Observations
1	The magnesium effervesces, disappears and the gas evolved extinguishes the lighted splint with a pop. On addition of sodium hydroxide solution, a white precipitate is formed.
2	There is an exothermic reaction and a brown solid can be seen. The solution formed is green in colour. The filtrate gives a green precipitate with sodium hydroxide solution.
3	The zinc disappears; the reaction is exothermic and the blue colour of the solution fades and a brown solid is formed. The colourless filtrate gives a white precipitate with sodium hydroxide solution and this redissolves on addition of excess sodium hydroxide solution.
4	The magnesium disappears and a black solid forms. The filtrate is colourless. When sodium hydroxide solution is added a white precipitate forms which does not dissolve on addition of more sodium hydroxide solution.

Table 4.1

Answers to the workbook questions (using the sample results)

Reaction 1

a The gas is hydrogen as shown by the lighted splint being extinguished with a pop. The Mg^{2+} ion is formed as shown by the white precipitate of magnesium hydroxide formed with sodium hydroxide.

b Ionic equation:

$$Mg(s) + 2H^+(aq) \longrightarrow Mg^{2+}(aq) + H_2(g)$$
$$Mg^{2+}(aq) + 2OH^-(aq) \longrightarrow Mg(OH)_2(s)$$

c This is a redox reaction because the magnesium is oxidised. Its oxidation state increases from 0 to +2 whilst the oxidation number of the hydrogen decreases from +1 to 0 showing it has been reduced.

Reaction 2

d One product is copper: the red–brown solid formed. The other product is the $Fe^{2+}(aq)$ ion. This is shown by the green solution formed and the green precipitate when the solution is reacted with sodium hydroxide solution.

e Ionic equation:

$$Cu^{2+}(aq) + Fe(OH)_2(s) \longrightarrow Cu(s) + Fe^{2+}(aq);$$
$$Fe^{2+}(aq) + 2OH^-(aq) \longrightarrow Fe(OH)_2(s)$$

f The copper(II) ion is reduced because its oxidation number decreases from +2 to 0. The iron is oxidised because its oxidation number increases from 0 to +2.

Reaction 3

g The zinc forms the Zn^{2+} ion. This is shown by the fact that when sodium hydroxide solution is added to the solution produced by the reaction, a white precipitate is formed ($Zn(OH)_2(s)$) which dissolves in excess sodium hydroxide solution. The other product is copper metal which is the brown solid formed. The $Cu^{2+}(aq)$ ion disappears as shown by the fading of the blue colour.

h $Zn(s) + Cu^{2+}(aq) \longrightarrow Zn^{2+}(aq) + Cu(s)$

$Zn^{2+}(aq) + 2OH^-(aq) \longrightarrow Zn(OH)_2(s);$

$Zn(OH)_2(s) + 2OH^-(aq) \longrightarrow Zn(OH)_4^{2-}(aq)$

i The copper(II) ion is reduced because its oxidation number decreases from +2 to 0. The zinc is oxidised because its oxidation number increases from 0 to +2.

> **Reaction 4**
>
> **j** The product contains the Mg^{2+} ion because this ion forms a white precipitate of magnesium hydroxide with sodium hydroxide which does not dissolve on addition of excess sodium hydroxide. This means that the black solid formed is zinc.
>
> **k** $Mg(s) + Zn^{2+}(aq) \longrightarrow Mg^{2+}(aq) + Zn(s)$
>
> **l** The magnesium is oxidised because its oxidation number increases from 0 to +2. The zinc ion is reduced because its oxidation number decreases from +2 to 0.

Practical investigation 4.2:
Understanding redox (II): investigating further reactions

Skills focus
The following skill areas are developed and practised (refer the skills grids at the front of this guide for codes):

MMO Collection of data and observations (b), (c) and (e)
      Decisions relating to measurements of observations (a), (d) (e) and (f)

ACE   Drawing conclusions (c) and (d)

Duration
The practical should take about one hour to complete.

Equipment
Each learner or group will need:
- 10 test tubes
- two test-tube racks
- six droppers (graduated if possible)
- plastic gloves
- small spatula

Access to:
- $0.02 \, mol \, dm^{-3}$ potassium manganate(VII) solution
- $0.10 \, mol \, dm^{-3}$ iron(II) sulfate solution
- $2 \, mol \, dm^{-3}$ sodium hydroxide solution
- '20 volume' hydrogen peroxide solution
- $1 \, mol \, dm^{-3}$ sulfuric acid
- $0.10 \, mol \, dm^{-3}$ sodium sulfite (sodium sulfate(IV)) solution.
- $0.10 \, mol \, dm^{-3}$ iron(III) sulfate solution
- $0.10 \, mol \, dm^{-3}$ barium chloride solution
- $0.10 \, mol \, dm^{-3}$ sodium sulfate (sodium sulfate(VI)) solution
- $2.0 \, mol \, dm^{-3}$ hydrochloric acid
- 1:1 hydrochloric acid (solution of equal volumes of concentrated hydrochloric acid and distilled water)

Safety considerations
- Learners must wear eye protection at all times.
- The potassium manganate(VII) solution is harmful and can cause brown stains on skin and clothing – if possible wear plastic gloves.
- '20 volume' hydrogen peroxide solution is an irritant and can cause burns and white stains on skin.
- The moderately concentrated hydrochloric acid is corrosive.

Carrying out the investigation
- The practical work is very simple but instead of just accepting the identities of the products formed, the learners are asked to confirm their conclusions by testing for different ions.
- The sulfite ion is very easily oxidised to sulfate and therefore when tested with barium chloride solution followed by hydrochloric acid there will be some barium sulfate formed giving a faint precipitate after the addition of hydrochloric acid. Therefore, the learners must be aware that some results are not as clear-cut as the theory suggests.
- The reaction between Fe^{2+} ions and hydrogen peroxide is a complex reaction but is treated simply here.
  ⓘ Learners could be asked to make up their own reaction mixture along with the chemical test to identify the ion formed in the reaction.

Common learner misconceptions
- If learners understand the concept of oxidation and reduction in terms of changes in oxidation number or by loss and gain of electrons then there is very little to misunderstand about the practical.

Sample results

Reaction number	Reactants	Observations corresponding to the instructions
1	Fe^{2+}(aq) and acidified MnO_4^-(aq)	**a** The potassium manganate (VII) is decolourised as soon as it is added to the iron(II) sulfate solution. **b** A brown precipitate is formed.
2	H_2O_2 and SO_3^{2-}(aq)	**a** A white precipitate is formed that does not dissolve on addition of the dilute acid. **b** A white precipitate is formed that dissolves on addition of the dilute acid. **c** When the barium chloride solution is added, a white precipitate is formed that does not dissolve on addition of the dilute acid.
3	Concentrated HCl and iron; addition of hydrogen peroxide solution to the product.	**a** There is a slow effervescence and a green solution is formed. **b** When sodium hydroxide solution is added to the first portion, a green precipitate is formed. When hydrogen peroxide solution is added to the second portion, an orange–brown solution is formed which gives a brown precipitate with sodium hydroxide solution.

Table 4.2

Answers to the workbook questions (using the sample results)

Reaction 1

a A reaction is shown by the fact that the potassium manganate(VII) is decolourised.

b The reaction is reduction because the oxidation number of the manganese decreases from +7 to +2.

c The half-equation for the oxidation of Fe^{2+} is $Fe^{2+} \longrightarrow Fe^{3+} + e^-$.

d $MnO_4^-(aq) + 8H^+(aq) + 5Fe^{2+}(aq) \longrightarrow Mn^{2+}(aq) + 4H_2O(aq) + 5Fe^{3+}(aq)$

This is a redox reaction because the oxidation state of the manganese has decreased from +7 to +2 and the manganese is therefore reduced. The oxidation state of the iron increases from +2 to +3 and it is therefore oxidised. This is shown by the formation of the brown precipitate with sodium hydroxide solution. $Fe^{3+}(aq) + 3OH^-(aq) \longrightarrow Fe(OH)_3(s)$

Reaction 2

e Sulfate ions form a (dense) white precipitate with Ba^{2+} ions and this precipitate does not react with acids. Sulfite ions also form a white precipitate with Ba^{2+} ions but this precipitate reacts with H^+ ions and dissolves.

$Ba^{2+}(aq) + SO_4^{2-}(aq) \longrightarrow BaSO_4(s)$

$Ba^{2+}(aq) + SO_3^{2-}(aq) \longrightarrow BaSO_3(s)$

$BaSO_3(s) + 2H^+(aq) \longrightarrow Ba^{2+}(aq) + SO_2(g) + H_2O(l)$

f The hydrogen peroxide oxidises the sulfite to sulfate. This can be seen from the fact that the precipitate formed with barium ions does not dissolve when acid is added.

$H_2O_2(aq) + SO_3^{2-}(aq) \longrightarrow SO_4^{2-}(aq) + H_2O(l)$

$Ba^{2+}(aq) + SO_4^{2-}(aq) \longrightarrow BaSO_4(s)$

g The oxidation number/state of the sulfur in the SO_3^{2-} ion is +4. In the SO_4^{2-} ion its oxidation number is +6. Therefore, the sulfur has been oxidised as its oxidation number has increased. In the hydrogen peroxide the oxidation number of the oxygen is −1; in water the oxidation number of the oxygen is −2. Therefore, the oxidation number of the oxygen has decreased and oxygen has been reduced.

Reaction 3a – the reaction between iron and hydrochloric acid.

h The effervescence is due to hydrogen. The green precipitate with sodium hydroxide solution shows that the ion present is Fe^{2+}(aq).

$Fe(s) + 2HCl(aq) \longrightarrow FeCl_2(aq) + H_2(g)$

OR $Fe(s) + 2H^+(aq) \longrightarrow Fe^{2+}(aq) + H_2(g)$

i The oxidation number of the iron has increased from 0 in the element to +2 in the product and it has been oxidised. The oxidation number of the hydrogen has decreased from +1 in the acid to 0 in the element. It has been reduced.

Reaction 3b – the reaction between the product of reaction 3a and hydrogen peroxide.

j   The presence of $Fe^{2+}(aq)$ in the solution is confirmed by the green precipitate with sodium hydroxide solution. $Fe^{2+}(aq) + 2OH^-(aq) \longrightarrow Fe(OH)_2(s)$

The hydrogen peroxide oxidises the $Fe^{2+}(aq)$ to $Fe^{3+}(aq)$. This is confirmed by the brown precipitate with sodium hydroxide solution.

$Fe^{3+}(aq) + 3OH^-(aq) \longrightarrow Fe(OH)_3(s)$

The equation for the reaction with hydrogen peroxide is as follows:

$2Fe^{2+}(aq) + H_2O_2(aq) + 2H^+(aq) \longrightarrow 2Fe^{3+}(aq) + 2H_2O(l)$

k   The iron(II) ions are oxidised from an oxidation state of +2 to an oxidation state of +3. The oxygen in the hydrogen peroxide is reduced from an oxidation state of −1 to −2 in water, so this is a redox reaction.

Chemical equilibrium

Chapter outline

This chapter refers to Chapter 8: Equilibrium in the coursebook.

In this chapter learners will complete practical investigations on:

- 5.1 Applying Le Chatelier's principle to a gaseous equilibrium
- 5.2 Applying Le Chatelier's principle to an aqueous equilibrium
- 5.3 The equilibrium constant for the hydrolysis of ethyl ethanoate.

Preparing for the investigations

- Learners should be familiar with the concept of a chemical equilibrium and the application of Le Chatelier's principle to both predict and explain the results of a change to equilibrium conditions.

- Learners should be able to predict the effects of pressure changes on gaseous equilibria and the effects of temperature on both gaseous and aqueous equilibria.

- Learners should also be able to interpret the signs of the enthalpy changes for both directions of an equilibrium from the effect of temperature changes on the equilibrium position.

- They will need to have some experience of how to calculate the equilibrium constant for a system at equilibrium. Practical investigation 5.3 should not be the first encounter learners have with this type of calculation.

- It is important that before completing the titrations required in Practical investigation 5.3, learners should be able to recognise that there will be an increase in the amount of acid formed when the initial concentration of ethyl ethanoate is increased.

Practical investigation 5.1:
Applying Le Chatelier's principle to a gaseous equilibrium

Introduction

In this investigation, learners will be applying Le Chatelier's principle to explain observations made on the reversible reaction between nitrogen dioxide and dinitrogen tetroxide:

$$2NO_2(g) \rightleftharpoons N_2O_4(g)$$

The nitrogen dioxide (NO_2) is a brown gas and the dinitrogen tetroxide (N_2O_4) is colourless. This means that any shift in the equilibrium position can be deduced from the intensity of the brown colour of the gaseous mixture. The greater the intensity of the brown colour, the more the equilibrium lies towards the left of the equation.

Skills focus

The following skill areas are developed and practised (refer the skills grids at the front of this guide for codes):

MMO	Collection of data and observations (a), (b) and (c)
	Decisions relating to measurements of observations (a), (c) and (d)
PDO	Recording data and observations (a) and (e)
	Display of calculations and reasoning (a) and (b)
	Data layout (b), (c), (d), (e) and (f)
ACE	Data interpretation and sources of error (a) and (b)
	Drawing conclusions (a), (b), (c) and (d)

Duration

- This practical will take about half an hour to complete.

Preparing for the investigation

- The main problems are outlined in the safety considerations

Equipment

Each learner or group will need:

- a 250 cm^3 beaker and a 100 cm^3 beaker
- some mouldable putty / modelling clay

- a plastic dropper with the narrow stem removed (see Figure 5.1)

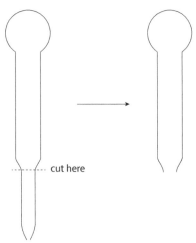

Figure 5.1

- a plain piece of white paper to use as a background for viewing colour changes
- beaker tongs (if a Bunsen burner is used for heating up water)

Access to:
- ice
- hot water (care) from a kettle or use a Bunsen burner
- copper turnings
- concentrated nitric acid
- dilute sodium hydroxide solution (e.g. 0.100 mol dm^{-3}) for disposal of the nitrogen dioxide at the end of the practical

Safety considerations

- Eye protection must be worn at all times during the investigation.
- Nitrogen dioxide is a toxic gas and all the operations involving it before it is in the sealed dropper **must** be done in a fume cupboard. The nitrogen dioxide can be produced in a boiling tube. Simply add the concentrated nitric acid to some copper turnings. The fume cupboard must be switched on at this stage.
- Concentrated nitric acid is corrosive and will give yellow stains on skin. Plastic gloves must be worn when dealing with the acid. It is probably better if the teacher fills up each dropper and secures the end with the chosen sealant material, for example, modelling clay.
- At the end of the practical work the nitrogen dioxide can be disposed of by bubbling it through dilute sodium hydroxide solution.
- Care must be taken when using the hot water.
- When carrying out the pressure investigation, each learner must make sure that they hold the sealant securely in the end of the dropper.

Carrying out the investigation

- The pressure investigation is the one that needs the closest scrutiny to ascertain what changes have taken place. The plain paper will aid this.

More able learners could be asked:

- To suggest a method that does not rely on their senses.
  They could use colorimetry where the instrument measures the intensity of the brown colour of the nitrogen dioxide.
- How does the sodium hydroxide remove the nitrogen dioxide? [HINT: sodium nitrite and sodium nitrate are formed].
  The nitrogen dioxide is an acidic oxide and it is neutralised by the alkali.
- To write the equation for the reaction.
  The reaction: $2NaOH(aq) + 2NO_2(g) \longrightarrow NaNO_2(aq) + NaNO_3(aq) + H_2O(l)$

Common learner misconceptions

- When dealing with the effects of pressure, learners sometimes get confused as to what constitutes the number of gas molecules and think that there is no change because the number of atoms does not change.
- Le Chatelier's principle sometimes causes problems when temperature changes are considered. If the temperature of the system is increased, what does the system do?

  It tries to lower the temperature by favouring the endothermic reaction. Therefore, the thermochemical nature of the forward and backward reactions can be ascertained from the effect of heating and cooling.

Sample results

Conditions		Observations
Effect of changing temperature	Increase	The gas turns a darker brown
	Decrease	The gas gets lighter in colour
Effect of changing pressure	Increase	The gas turns lighter in colour

Table 5.1

> **Answers to the workbook questions (using the sample results)**
>
> **a i** When the temperature is decreased, the amount/concentration/partial pressure of nitrogen dioxide decreases.
>
>   **ii** When the temperature is increased, the amount/concentration/partial pressure of nitrogen dioxide increases.
>
>   **iii** An increase in temperature favours the formation of nitrogen dioxide. Therefore, its formation (the backward reaction) must be an endothermic reaction because this resists the increase in temperature (Le Chatelier's principle). If the temperature is decreased, the exothermic reaction is favoured because this would raise the temperature and counteract the decrease in temperature. Therefore, the forward reaction is exothermic.
>
> **b** When the pressure is increased, the amount of nitrogen dioxide decreases. This is because an increase in pressure favours the reaction that leads to a decrease in the number of gas molecules and therefore more colourless dinitrogen tetroxide is formed as the equilibrium shifts in favour of the forward reaction.

Practical investigation 5.2:
Applying Le Chatelier's principle to an aqueous equilibrium

Introduction
In this investigation, the learners will apply Le Chatelier's principle to the aqueous equilibrium shown below:

$$[Cu(H_2O)_6]^{2+}(aq) + 4Cl^-(aq) \rightleftharpoons [CuCl_4]^{2-}(aq) + 6H_2O(l)$$
Blue — Yellow

Skills focus
The following skill areas are developed and practised (refer the skills grids at the front of this guide for codes):

MMO Collection of data and observations (a), (b), (c) and (e)
Decisions relating to measurements of observations (c) and (d)

PDO Recording data and observations (a) and (e)

ACE Data interpretation and sources of error (a) and (b)
Drawing conclusions (a) and (b)

Duration
- This practical will take three quarters of an hour to complete.

Equipment
Each learner or group will need:
- one plastic dropper
- 10 test tubes
- two boiling tubes (one for concentrated hydrochloric acid and one for copper(II) sulfate solution)

Cambridge International AS & A Level Chemistry

- one rubber bung that fits boiling tube
- 100 cm³ beaker for distilled water
- permanent marker pen
- test-tube rack that can accommodate at least two boiling tubes
- distilled water in a wash bottle
- three 250 cm³ beakers
- a sheet of plain white paper to act as a background

Access to:
- concentrated hydrochloric acid
- 1 mol dm⁻³ aqueous copper(II) sulfate
- distilled water
- ice

Safety considerations

- Eye protection must be worn at all times during the investigation.
- Concentrated hydrochloric acid is corrosive and the fumes escaping into the lab can be minimised by making sure the bung for the boiling tube is in place in the boiling tube.
- Copper(II) sulfate is a harmful environmental hazard and should be handled carefully.

Carrying out the investigation

- The size of the drops added to each reaction mixture must be consistent. That is why only one dropper should be used. After the aqueous copper(II) ions have been added, the same dropper must be washed thoroughly using distilled water and then concentrated hydrochloric acid.
- When the learners carry out the second part of the experiment on the effect of temperature, the difference in colour between the reaction mixtures at 0 °C and room temperature is quite small; if the corresponding test tubes are removed from the beakers, the difference becomes even harder to appreciate.
- ⓓ After they have carried out the experiments, learners could be asked: Why is the reaction between Cl⁻(aq) ions and $[Cu(H_2O)_6]^{2+}$(aq) exothermic?

The negatively charged Cl⁻(aq) ions are more strongly attracted to the Cu^{2+} ions than the neutral water molecules, so stronger bonds are formed and the reaction is exothermic. NOTE: The entropy also increases because more water molecules are released than Cl⁻ ions are bound. This increase in entropy will also drive the reaction. However, entropy is not covered until Advanced level and its use at this level is problematic.

a Why is the reverse reaction endothermic?

When water molecules replace the Cl⁻ ions, the stronger bonds have to be broken and this requires the transfer of energy from the surroundings, so the reaction is endothermic.

b What extra experiment(s) could be done to add extra evidence to this thermochemical investigation?

One tube could be heated to 100 °C and another could be added to a freezing mixture of ice and sodium chloride, or ice and calcium chloride, where the temperature will be below 0 °C.

Common learner misconceptions

- Learners do not usually find this investigation difficult to follow; the colour change is easy to see and the trend is unmistakable.
- The effect of temperature is not difficult to interpret if Le Chatelier's principle is followed.

Sample results

Part 1 – The effect of concentration changes on the position of equilibrium

Trend in colour as the concentration of hydrochloric acid (Cl⁻ ions) is increased:

Blue ⟶ Blue–green ⟶ Green ⟶ Yellow–green ⟶ Yellow

Answers to the workbook questions (using the sample results)

Part 1

a As the concentration of Cl⁻ is increased then, according to Le Chatelier's principle, the equilibrium will shift to lower its concentration and counteract the change. It does this by favouring the forward reaction and therefore forming the yellow $[CuCl_4]^{2-}$ ion. The intermediate colours are due to mixtures of this yellow ion and the blue $[Cu(H_2O)_6]^{2+}$ ion.

Part 2 – The effect of temperature on the position of equilibrium

Conditions	Observations
0 °C	The mixture has a more yellow appearance
Room temperature	Control – yellow–green in colour
Boiling water	The mixture is a darker green in colour

Table 5.2

Chapter 5: Chemical equilibrium

> **Answers to the workbook questions (using the sample results)**
>
> **Part 2**
>
> **b i** It is increased in concentration
>
>   **ii** It is decreased in concentration
>
> **c** The forward reaction in which $[CuCl_4]^{2-}$ is formed must be exothermic because it is favoured by a decrease in temperature. In accordance with Le Chatelier's principle, an exothermic reaction will counteract the decrease in temperature.

Practical investigation 5.3:
Equilibrium constant for the hydrolysis of ethyl ethanoate

Introduction

In this investigation learners will determine the equilibrium constant (K_c) for the hydrolysis of ethyl ethanoate:

$$CH_3COOC_2H_5(l) + H_2O(l) \underset{}{\overset{H^+ \text{catalyst}}{\rightleftharpoons}} CH_3COOH(l) + C_2H_5OH(l)$$

Skills focus

The following skill areas are developed and practised (refer to skills grid section for codes):

MMO	Collection of data and observations (a), (b), (c), (d) and (e)
	Decisions relating to measurements of observations (a), (b) and (c)
PDO	Recording data and observations (a) and (e)
	Display of calculations and reasoning (a) and (b)
ACE	Data interpretation and sources of error (a), (b), (d), (g), (h) and (i)
	Drawing conclusions (a), (c) and (d)

Duration

- This practical will need a one-hour lesson to set it up and the experimental mixtures should be left for at least two days before the analysis is completed. The titrations should be completed on the same day.

Preparing for the investigation

- The investigation is in two parts and ideally these should be separated by an interval of at least 2 days to allow the mixture to come to equilibrium. The time between the two parts should be used to make sure that the learners are aware of the calculations and how a value for K_c is arrived at.

- To save on reagents, especially the ethyl ethanoate, different groups can be allocated different mixtures to investigate and the results pooled.

- The class should be working in groups and this facilitates the accurate transcription of the volumes of reagents from the instructions as there are quite a few different volumes to measure.

Equipment

Each learner or group will need:

- 500 cm^3 volumetric flask
- thymolphthalein indicator
- 250 cm^3 conical flask
- white tile
- small filter funnel for filling burette
- six sample tubes which can take up to 10 cm^3 with tight fitting plastic tops. If these are not available then boiling tubes will do and these will require size 21 bungs to make them airtight
- container for sample tubes or rubber band to keep them together
- permanent marker pen
- 50.00 cm^3 burette
- two 5.00 cm^3 or 10.00 cm^3 graduated pipettes. One of these is for the ethyl ethanoate and the other is for the aqueous solutions
- wash bottle of distilled water

Access to:

- ethyl ethanoate
- 2.00 mol dm^{-3} hydrochloric acid
- 1.00 mol dm^{-3} sodium hydroxide solution

Safety considerations

- Eye protection must be worn at all times.
- The sodium hydroxide solution is corrosive at this concentration.
- Ethyl ethanoate is volatile, flammable and its vapour is harmful in an enclosed space.
- The dilute hydrochloric acid is an irritant at this concentration. Special care should be taken with the titration.

Carrying out the investigation

- Learners should not get their graduated pipettes mixed up because the ethyl ethanoate is immiscible with water and care should be taken not to contaminate it.
- The most difficult problem is the titration because learners do not have the luxury of being able to repeat the titrations and check their results. This is why an estimate is required for the first titration of the dilute hydrochloric acid. It gives them some idea of when they should start running in the alkali drop by drop from the burette. In fact, to be really careful they can do this 2 cm³ before they reach the estimated volume.
- Learners should also be aware that as the initial ethyl ethanoate concentration is increased, the volume of alkali required will increase and they should understand why this is so. This understanding will help them have some idea of the volume of alkali required for each experiment.

Very able students can be asked to make up their own automated spreadsheet using Microsoft Excel or a similar programme.

Common learner misconceptions

- Learners do not automatically assume that water is part of the reaction mixture. They will need to be reminded that it should be included in the expression for K_c and that it is the major component of dilute hydrochloric acid compared with HCl. They will also be surprised at the high concentration of HCl compared with the other reactants and products.
- Another quantity that can cause learners problems is the magnitude of the concentration of water. Learners are not used to concentrations of substances being 20 and 50 mol dm⁻³. Note: it is assumed that the hydrochloric acid is mostly water.

Sample results

Part 2: Analysis of the reaction mixtures to determine the composition of the equilibrium mixture

Table 5.3 shows sample results for the titration of the standard sodium hydroxide solution against the equilibrium mixture resulting from the hydrolysis of ethyl ethanoate. There are two sets of results for each combination of reactants. This gives some idea of the variation in titration values for each set of results.

Tube	1	1a	2	2a	3	3a	4	4a	5	5a	6	6a
Final burette reading/cm³	9.70	19.70	39.05	18.95	46.00	27.20	33.40	34.70	38.85	39.30	41.65	42.10
Initial burette reading/cm³	0.00	10.00	20.00	0.00	18.95	0.00	0.00	0.90	0.95	1.30	0.20	0.75
Titre/cm³	9.70	9.70	19.05	18.95	27.05	27.20	33.40	33.80	37.90	38.00	41.45	41.35

Table 5.3

Chapter 5: Chemical equilibrium

Answers to the workbook questions (using the sample results)

a As the concentration/amount of ethyl ethanoate increases, the volume of alkali increases. This shows that the concentration of ethanoic acid also increases.

b
i If the concentration of ethyl ethanoate is increased, the equilibrium will shift so as to reduce it. It does this by favouring the forward reaction, which means a greater concentration of ethanoic acid.

ii The expression for K_c is $[CH_3COOH][C_2H_5OH]/[CH_3COOC_2H_5][H_2O]$. If $[CH_3COOC_2H_5]$ is increased then the value for K_c will be decreased in value. At constant temperature K_c is constant. Therefore, to return K_c to its real value, the concentrations of ethanoic acid and ethanol must increase and the concentrations of water and ethyl ethanoate must decrease. This explains why the $[CH_3COOH]$ increases and the volume of alkali needed to neutralise it also increases.

Worked example using sample results for Tube 2

c The volume of alkali required by control Tube 2 = 9.70 cm³.

d $[CH_3COOH]_{eqm}$

i Extra volume of alkali = 19.05 − 9.70 = 9.35 cm³

ii Number of moles of CH_3COOH at equilibrium
= No. of moles of ethanoic acid = number of extra moles of alkali required
= $1.00 \times 9.35 \times 10^{-3}$ mol

iii Equilibrium concentration of CH_3COOH
$= \dfrac{n}{V} = \dfrac{9.35 \times 10^{-3}}{10 \times 10^{-3}} = 0.935$ mol dm⁻³

e $[C_2H_5OH]_{eqm.}$
$[C_2H_5OH]_{eqm} = \dfrac{n}{V} = 0.935$ mol dm⁻³

f Equilibrium concentration of ethyl ethanoate

i Initial number of moles of $CH_3COOC_2H_5$
Mass of $CH_3COOC_2H_5$ initially = density × volume
= 0.900 × 1.00
= 0.900 g

ii Number of moles of $CH_3COOC_2H_5$ that have reacted
Initial number of moles = $\dfrac{mass}{Mr}$
$= \dfrac{0.9}{88}$ mol
= 0.01023 mol

iii The number of moles of $CH_3COOC_2H_5$ that react = the number of moles of ethanoic acid that are formed
$= 9.35 \times 10^{-3}$ mol

iv Number of moles of $CH_3COOC_2H_5$ at equilibrium
= initial number − number that reacted
$= 0.01023 - 9.35 \times 10^{-3}$ mol

v Equilibrium concentration of $CH_3COOC_2H_5$
$= \dfrac{n}{V} = \dfrac{8.8 \times 10^{-4}}{10 \times 10^{-3}}$ mol dm⁻³
$= 8.8 \times 10^{-2}$ mol dm⁻³

g Calculate the equilibrium concentration of water

i Initial mass of water = density × volume
Volume of water = 9.00 cm³
Mass of water = 1 × 9 = 9.00 g

ii Initial number of moles of water = $\dfrac{m}{Mr} = \dfrac{9}{18} =$ 0.500 mol

iii Number of moles of water that react = number of moles of $CH_3COOC_2H_5$ that react
$= 9.35 \times 10^{-3}$

iv Number of moles of water at equilibrium = initial number − number that reacted
$= 0.500 - 9.35 \times 10^{-3} = 0.4907$ mol

v ∴ Equilibrium concentration of water =
$\dfrac{n}{V} = \dfrac{0.4907}{10 \times 10^{-3}} = 49.07$ mol dm⁻³

h Expression for K_c
$K_c = \dfrac{[\text{ethanoic acid}][\text{ethanol}]}{[\text{ethyl ethanoate}][\text{water}]}$

i Value of K_c in this experiment
$K_c = \dfrac{0.935 \times 0.935}{0.088 \times 49.07} = 0.202$ (no units)

Calculations for Tubes 3–6 based on sample results

j Tube 3

Extra volume of NaOH required/ cm^3	No. of moles of ethanoic acid at eqm/ mol	[CH$_3$COOH] equilibrium mol dm^{-3}	[C$_2$H$_5$OH] at eqm/ mol dm^{-3}	Initial no. of moles of ester/ mol	No. of moles of ester at eqm/ mol	[ester] at eqm/ mol dm^{-3}	Initial no. of moles of water / mol	No. of moles of water at eqm/ mol	[water] at eqm/ mol dm^{-3}
17.35	1.735 × 10^{-2}	1.735	1.735	0.02045	3.1 × 10^{-3}	0.3100	0.4444	0.4271	42.71

$$K_c = \frac{1.735 \times 1.735}{0.31 \times 42.71} = 0.227 \text{ (no units)}$$

k Tube 4

Extra volume of NaOH required /cm^3	No. of moles of ethanoic acid at eqm./ mol	[CH$_3$COOH] at equilibrium mol dm^{-3}	[C$_2$H$_5$OH] at eqm.	Initial no. of moles of ester	No. of moles of ester at eqm.	[ester] at eqm.	Initial no. of moles of water	No. of moles of water at eqm.	[water] at eqm.
23.70	0.0237	2.370	2.370	0.03068	6.982 × 10^{-3}	0.6982	0.3889	0.3652	36.52

$$K_c = \frac{2.37 \times 2.37}{0.6982 \times 36.52} = 0.220 \text{ (no units)}$$

l Tube 5

Extra volume of NaOH required /cm^3	No. of moles of ethanoic acid at eqm./ mol	[CH$_3$COOH] at equilibrium mol dm^{-3}	[C$_2$H$_5$OH] at eqm.	Initial no. of moles of ester	No. of moles of ester at eqm.	[ester] at eqm.	Initial no. of moles of water	No. of moles of water at eqm.	[water] at eqm.
28.20	0.0282	2.82	2.82	0.04091	0.0127	1.27	0.3333	0.3051	30.51

$$K_c = \frac{2.82 \times 2.82}{1.27 \times 30.51} = 0.205 \text{ (no units)}$$

m Tube 6

Extra volume of NaOH required/ cm^3	No. of moles of ethanoic acid at eqm/ mol	[CH$_3$COOH] equilibrium mol dm^{-3}	[C$_2$H$_5$OH] at eqm/ mol dm^{-3}	Initial no. of moles of ester/ mol	No. of moles of ester at eqm/ mol	[ester] at eqm/ mol dm^{-3}	Initial no. of moles of water / mol	No. of moles of water at eqm/ mol	[water] at eqm/ mol dm^{-3}
31.75	0.03175	3.175	3.175	0.05114	0.01.939	1.939	0.2778	0.2460	24.60

$$K_c = \frac{3.175 \times 3.175}{1.939 \times 24.60} = 0.211 \text{ (no units)}$$

n All the results fall into the category for the accepted value for K_c

o Average value for K_c = 0.222

p Percentage error = $\frac{((0.222-0.220) \times 100)}{0.220}$ = 3.18%

Chapter 6:
Rates of reaction

Chapter outline

This chapter refers to Chapter 9: Rates of reaction in the coursebook.

In this chapter learners will complete investigations on:

- 6.1 Effects of concentration on rate of chemical reaction
- 6.2 Effects of temperature and a homogeneous catalyst on the rate of chemical reaction
- 6.3 Observe a heterogeneous catalysed reaction

Practical investigation 6.1:
Effects of concentration on rate of chemical reaction.

Skills focus

The following skill areas are developed and practised (refer the skills grids at the front of this guide for codes):

MMO	Collection of data and observations (a), (b), (c), (d) and (e)
	Decisions relating to measurements of observations (c)
PDO	Recording data and observations (a), (c) and (e)
	Display of calculations and reasoning (a) and (b)
	Data layout (a) (b), (c), (d), (e) and (f)
ACE	Data interpretation and sources of error (a), (b), (d), (e) and (f)
	Drawing conclusions (a), (b), (c) and (d)
	Suggesting improvements (a) and (b)

Duration

The practical will take about two hours to complete.

Preparing for the investigation

- This practical investigation is designed so that the concentration of the acid reactant is varied but the amount of acid is constant. This means that the total volume of gas evolved each time is constant.

- The practical is not unlike some of those that learners would have done at IGCSE or O Level but they are asked to analyse the results obtained to find initial rates in terms of $mol\,dm^{-3}\,s^{-1}$. These units are used in the A Level course

- The time intervals for measurement are designed to give the learners enough time to read the volume measurement, record it and then anticipate the next measurement.

Equipment

Each learner or group will need:

- one of the two sets of apparatus for measuring gaseous volumes (see Skills chapter)
- three conical flasks with a capacity of 150 cm³ or three boiling tubes with a capacity of 40 cm³
- weighing boat
- 10 cm³ graduated pipette for accurate measurement of hydrochloric acid volume
- wash bottle of distilled water
- dropper
- stopwatch

Access to:

- hydrochloric acid in three different concentrations: $0.500\,mol\,dm^{-3}$, $0.750\,mol\,dm^{-3}$ and $1.00\,mol\,dm^{-3}$.
- small marble chips (2–4 mm)
- a top-pan balance reading to at least two decimal places

Safety considerations

- Learners must wear eye protection at all times.
- The hydrochloric acid is an irritant at the concentrations in the experiment.

Carrying out the investigation

- When weighing out the small marble chips, the surface area should be as similar as possible for each experiment. Therefore, the same number of chips should be used each time.
- The selection of marble chips can take time. If small marble chips (<4 mm) are used then about seven will be required for a reasonable approximation of a constant surface area. The best solution may be to give each group a selection of marble chips from which to select their three samples of chips before they start the determinations. They do not need exactly 1.00 g. An amount between 0.95 and 1.05 g will suffice because the calcium carbonate is in excess.
- If the marble chips are coated with powder they can be washed with lots of water and dried in an oven before use.
- The conical flask used as a reaction container should be swirled continuously to ensure smooth evolution of gas.
- The volumes of acid required are quite precise so care should be taken to measure the volumes accurately using graduated pipettes.

⚙ Ask learners to predict the volume of carbon dioxide they should get in all three reactions.

- You can ask learners to measure the gradients at various points on one of the graphs and for each gradient, calculate the concentration of the acid at that point. For example, the volume of gas that could be evolved from each of the three experiments is 96 cm^3. Therefore, if we choose the gradient at 48 cm^3, then half of the acid has been used up and the concentration has been halved.

Common learner misconceptions

- It is unusual for learners to use different volumes of reactant in a rate determination. Some may expect to get smaller volumes of gas from the smaller volumes of reactant.
- Because different volumes are used, learners may also believe that the investigation is not a fair test.

Sample data

Experiment 1

Table 6.1: 16.0 cm^3 of 0.500 mol dm^{-3} hydrochloric acid

Time/s	0	15	30	45	60	75	90	105	120	135
Vol. of gas/cm^3	0.0	3.0	6.0	10.0	14.0	18.0	21.0	25.0	29.0	33.0
Time/s	150	165	180	195	210	225	240	255	270	285
Vol. of gas/cm^3	36.0	39.0	43.0	46.0	49.0	52.0	55.5	59.0	61.0	63.5
Time/s	300	330	360	390	420	450	480	510	540	570
Vol. of gas/cm^3	65.0	68.0	71.0	75.0	77.0	80.0	81.5	83.5	84.5	85.5

Table 6.1

Experiment 2

Table 6.2: 10.70 cm^3 of 0.750 mol dm^{-3} hydrochloric acid

Time/s	0	15	30	45	60	75	90	105	120	135
Vol. of gas/cm^3	0.0	6.0	11.0	15.5	22.0	26.0	31.0	35.0	40.0	44.0
Time/s	150	165	180	195	210	225	240	255	270	285
Vol. of gas/cm^3	48.0	51.0	54.0	57.5	60.0	63.0	65.5	67.5	70.0	71.5
Time/s	300	330	360	390	420	450	480	510	540	570
Vol. of gas/cm^3	73.5	77	79.5	80.5	82.0	83.5	84.5	85.0	85.5	86.0

Table 6.2

Chapter 6: Rates of reaction

Experiment 3

Table 6.3: 8.00 cm³ of 1.00 mol dm⁻³ hydrochloric acid

Time/s	0	15	30	45	60	75	90	105	120	135
Volume of gas/cm³	0.0	9.0	17.5	25.0	32.0	40.0	46.0	52.0	56.5	62.0
Time/s	150	165	180	195	210	225	240	255	270	285
Volume of gas/cm³	67.0	70.5	74.0	77.0	80.0	82.0	85.0	86.5	87.0	87.0
Time/s	300	330	360	390	420	450	480	510	540	570
Volume of gas/cm³	87.0	87.0	87.0							

Table 6.3

Answers to the workbook questions (using the sample results)

a Learner graphs should use at least three quarters of the graph paper and each line should be plotted using different colours or symbols. The final lines should be smooth curves.

b i As the concentration increases, the curve becomes steeper and the final volume of gas is reached in a shorter time.

ii There is some variation but the final volume will be in the region of 85–95 cm³. If learners' state that the final volume is the same for each concentration allowing for experimental error, then they are correct.

c Slopes of the tangent increase as the concentration is increased, showing that the initial rate is faster if the concentration is increased.

d See Table 6.4.

e See Figure 6.1

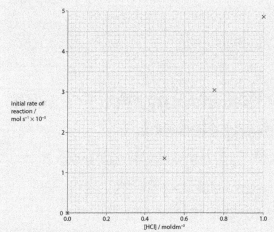

Figure 6.1

f 0,0 – because if there is no acid, there is no reaction and the rate is 0.

g As the concentration increases, the rate also increases but it is not directly proportional. It could be a curve.

h Carry out investigations of the rate at a lower concentration (e.g. 0.250 mol dm⁻³) and a higher concentration (e.g. 1.00 mol dm⁻³).

[HCl(aq)]/ mol dm⁻³	Rate/production of CO_2 cm³ min⁻¹	Rate – production of CO_2/cm³ s⁻¹	Rate – production of CO_2/mol s⁻¹	Rate – removal of HCl(aq)/ mol s⁻¹
0.500	10.5	$\frac{10}{60} = 0.175$	$\frac{0.166}{24000} = 6.94 \times 10^{-6}$	$6.94 \times 10^{-6} \times 2 = 1.38 \times 10^{-5}$
0.750	22.0	$\frac{22}{60} = 0.367$	$\frac{0.367}{24000} = 1.53 \times 10^{-5}$	$1.53 \times 10^{-5} \times 2 = 3.06 \times 10^{-5}$
1.00	34.0	$\frac{35}{60} = 0.583$	$\frac{0.583}{24000} = 2.430 \times 10^{-5}$	$2.360 \times 10^{-5} \times 2 = 4.86 \times 10^{-5}$

Table 6.4

i See Table 6.5.

Control variable	How it was kept constant
Temperature	Experiments done on same day in same place
Surface area of calcium carbonate	Same mass and approximately same number of lumps
Final volume of gas	Use equal number of moles of acid. Acid is the limiting reactant.

Table 6.5

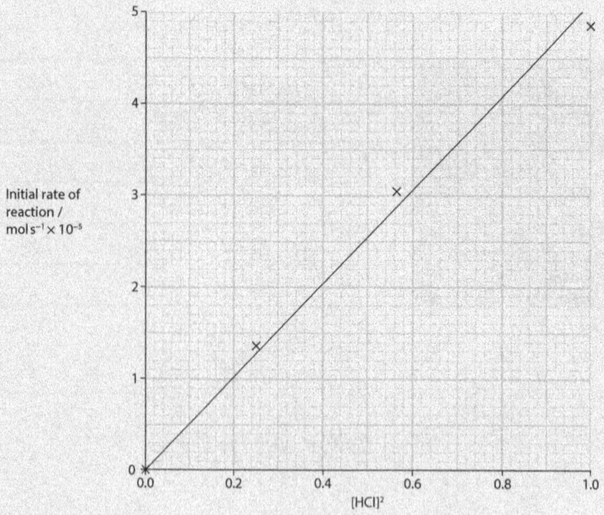

Figure 6.2

Practical investigation 6.2: with planning element
Effects of temperature and a homogeneous catalyst on the rate of chemical reaction

Skills focus

The following skill areas are developed and practised (refer the skills grids at the front of this guide for codes):

MMO Collection of data and observations (b), (c) and (e)

PDO Recording data and observations (c) and (d)

ACE Data interpretation and sources of error (a)
Drawing conclusions (a), (c) and (d)
Suggesting improvements (a), (b) and (c)

Duration

The practical will take three-quarters of an hour to complete.

Preparing for the investigation

- In this practical investigation, learners are asked to plan a very short investigation. This must not be elaborate and should require very little time.

Equipment

Each learner or group will need:

- six test tubes
- test-tube rack
- Bunsen burner and heat-resistant pad
- four droppers
- permanent marker pen
- anti-bumping granules

Access to:

- approximately $0.0200\ mol\ dm^{-3}$ potassium manganate(VII) solution
- $0.100\ mol\ dm^{-3}$ potassium or sodium ethanedioate solution
- $1\ mol\ dm^{-3}$ sulfuric acid
- $0.100\ mol\ dm^{-3}$ manganese(II) sulfate solution

Safety considerations

- Learners must wear eye protection at all times.
- The sulfuric acid is an irritant at the concentrations in the experiment.
- The potassium manganate(VII) solution will stain skin brown and should be handled with care.

Carrying out the investigation

- Some learners may believe that increasing the temperature 'lowers the activation energy'.
- Ask them to suggest ways of following the reaction in order to determine the reaction rate.

Sample data

Step 4 – the pink colour does not change when the potassium manganate(VII) is added.

Step 5 – after heating there may be a brown coloration but eventually the solution turns colourless showing the manganate(VII) ion has reacted.

Answers to the workbook questions (using the sample results)

a i Step 4 – the reaction between the manganate (VII) ion and the ethanedioate ion is slow because both ions are negatively charged and repel each other. This makes the activation energy large.

ii Step 5 – raising the temperature means that more particles have an energy greater than the activation energy. Therefore, more particles can react and the reaction is quicker.

b Use a colorimeter. This can be used because the intensity of the purple colour changes during the reaction.

c Learner methods should include the following:

- Add a few drops of the aqueous manganese (II) ions to the ethanedioate solution and mix thoroughly.
- Add the manganate(VII) solution to the 'cold' solution and observe.

d There is an immediate reaction without heating.

e The $Mn^{2+}(aq)$ ions are catalysts for the reaction and therefore lower the activation energy. This means that more particles have enough energy to react without raising the temperature and reaction takes place immediately.

Practical investigation 6.3: Demonstration
Observing a catalysed reaction

Introduction
As this investigation is designed to be a demonstration; the chemicals and apparatus are not listed in the learner's workbook.

Skills focus
The following skill areas are developed and practised (refer the skills grids at the front of this guide for codes):

MMO   Collection of data and observations (e)

ACE   Drawing conclusions (a), (c) and (d)

Duration
This practical will need a quarter of an hour.

Preparing for the investigation

- This practical demonstration requires a fume cupboard and, if possible, a room that can be made dark in order to facilitate the observations of the changes taking place.

Equipment
You will need:

- a small conical flask – no bigger than 100 cm^3. Any larger and the ammonia concentration may be insufficient for reaction to take place.
- stopper for conical flask
- Bunsen burner and heat-resistant pad
- glass rod
- a section of platinum wire – these are not very expensive and can obviously be used again and again.
- concentrated (880) ammonia fresh from the reagent bottle

Safety considerations

- You must have access to a fume cupboard. The ammonia fumes are very strong and can be overpowering.
- The class should be seated or standing at least 3 m from the fume cupboard and the fans should be on at all times.

- Never hold a bottle of concentrated ammonia by the neck only. Always support the base with one hand and use the other hand to hold the neck. If possible use a designated reagent bottle carrier/holder when carrying the bottle further than a few metres.

- Wear safety goggles for the experiment. The learners should be wearing safety spectacles even though they are not participating.

Carrying out the investigation

Suggested method for demonstration

1. Place a few cm³ of the concentrated ammonia solution in the flask and stopper the flask. Keep the flask away from the Bunsen burner so it is relatively cool.

2. Make a coil of the wire and suspend it from the glass rod into the flask as shown in Figure 6.3.

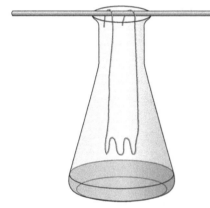

Figure 6.3

3. Take the wire out and heat strongly for a few seconds. In the darkened room the learners should see it glow and then not glow after it is removed from the heat.

4. Now place the wire back into the flask. Observe any changes both to the wire and what is present in the flask.

5. Remove from the flask and observe any changes.

6. Repeat Steps 4 and 5 a few times.

7. If the reaction ceases to happen then simply heat the wire strongly for a few seconds and resume.

Answers to the workbook questions (using the sample results)

a Learner observations:
- There is nothing changing inside the flask before the wire is placed inside it
- The wire glows red hot when it is in the flask
- It ceases to glow when it is removed from the flask
- The contents of the flask become misty and there may even be a slight brown coloration.

b Evidence of reaction:
- The fumes inside the flask
- The exothermic reaction taking place as shown by the red-hot wire.

c Heterogeneous – the solid platinum wire is in a different state to the reactants and products, which are gases.

d The wire glows red hot showing a reaction is taking place on the surface of the platinum.

Chapter 7:
The properties of metals

Chapter outline

This chapter relates to Chapter 10: Periodicity and Chapter 11: Group 2 in the coursebook.

In this chapter learners will complete investigations on:

- 7.1 Properties of metal oxides and metal chlorides across Period 3
- 7.2 Relative atomic mass of magnesium using a back-titration method
- 7.3 Planning investigation into the separation of two metal ions in solution
- 7.4 Identification of three metal compounds using qualitative analysis

Practical Investigation 7.1:
Properties of metal oxides and metal chlorides across Period 3

Introduction
This practical is designed to support theory work regarding periodicity. The concept could be extended to address the whole period, investigating the oxides and chlorides of the non-metals.

Skills focus
The following skill areas are developed to practised (refer the skills grids at the front of this guide for codes):

MMO	Collection of data and observations (a), (b), (c), (d) and (e)
	Decisions relating to measurements of observations (c)
PDO	Recording data and observations (a), (c) and (e)
	Display of calculations and reasoning (a) and (b)
	Data layout (a) (b), (c), (d), (e) and (f)
ACE	Data interpretation and sources of error (a), (b), (d), (e) and (f)
	Drawing conclusions (a), (b), (c) and (d)
	Suggesting improvements (a) and (b)

Duration
This practical will take one hour to complete.

Preparing for the investigation
- This investigation should ideally be completed when learners are covering periodicity, so that the concepts in it are very prominent in their minds.
- Learners should be aware of what is meant by hydrolysis and that polar covalent bonds are affected by it.
- All chlorides and oxides should be kept in a desiccator before the experiment is carried out.
- As sodium oxide is very difficult to obtain, a solution of sodium hydroxide is a good compromise.

Equipment
- test tubes and test-tube rack
- a dropper
- Universal Indicator (U.I.) paper
- eye protection
- small spatula (e.g. Nuffield type)
- wash bottle filled with distilled water
- a dropper bottle of Universal Indicator

Access to:
- a solution of sodium oxide
- solid magnesium oxide
- solid aluminium oxide
- anhydrous magnesium chloride
- anhydrous aluminium chloride
- sodium chloride

Safety considerations

- Sodium hydroxide is corrosive.
- The hydrogen chloride evolved during the hydrolysis of anhydrous aluminium chloride gives misty fumes of hydrochloric acid in the laboratory, but most of it is absorbed by the water that reacts with the aluminium chloride.
- The Universal Indicator solution is made up in ethanol which is a flammable liquid.

Carrying out the investigation

- The biggest problem is contamination of the metal chlorides with water. Anhydrous magnesium chloride is difficult to obtain completely dry. When added to water there is an exothermic reaction.
- The reaction between anhydrous aluminium chloride and water can be quite violent and therefore care must be taken when carrying out this reaction.

  🔍 Diagrams of metal–chlorine bonds with partial charges and the attractions of the two parts of the water molecule for the opposite ends of the bond will help in deciphering what happens when the H–OH bond splits.

  ⚙ Ask leaners to apply what they have learned in this exercise to Period 4 and predict what happens when water is added to the oxides and chlorides of potassium, calcium and gallium. Alternatively, they can predict the properties of the non-metal oxides and chlorides in the rest of Period 3. Their predictions could then be tested.

Common learner misconceptions

- Many learners have difficulty in understanding the concept of hydrolysis.
- Learners may apply the rule (that the –OH group attaches itself to the electron deficient atom and the hydrogen attaches itself to the electronegative atom) to **all** the metal chlorides. So, for example, sodium chloride reacts with water to give sodium hydroxide and hydrogen chloride. They therefore need to be reminded that hydrolysis will only occur where there is some form of covalent character in the metal–chlorine bond.

Sample results

Part 1: Testing metal oxides

See Table 7.1

Test tube	Observations	Conclusions
i Na_2O	U.I. turns purple	Sodium oxide is strong alkali
ii MgO	U.I. turns blue	Magnesium oxide is a weak(er) alkali
iii Al_2O_3	No change	Aluminium oxide does not produce OH^- ions in water

Table 7.1

Part 2: Testing metal chlorides

See Table 7.2

Test tube	Observations	Conclusions
i NaCl	U.I. goes green and there is very little temperature change	NaCl is a neutral chloride and simply dissolves in water to give a neutral solution
ii $MgCl_2$	U.I. goes green. There is an exothermic change. **Alternatively,** if the $MgCl_2$ is really dry then there will be a very exothermic change and the U.I. turns yellow-green	$MgCl_2$ dissolves in water to give a neutral solution. The heat change is due to hydration of the Mg^{2+} ion. **Alternatively,** there is partial hydrolysis and HCl forms to give a very dilute acidic solution
iii $AlCl_3$	U.I. goes red and there is an exothermic reaction. Misty fumes are formed which turn U.I. paper red. Faint white precipitate possibly formed.	Misty acidic fumes show HCl gas is formed. Faint white precipitate is aluminium hydroxide.

Table 7.2

Chapter 7: The properties of metals

> ## Answers to the workbook questions (using the sample results)
>
> ### Part 1
>
> **a** They get less alkaline as you go across the Period from left to right, starting off as strong alkalis and then getting steadily weaker.
>
> **b i** Sodium oxide $Na_2O(s) + H_2O(l) \rightarrow 2NaOH(aq)$ or $O^{2-}(s) + H_2O(l) \rightarrow 2OH^-(aq)$
>
> **ii** Magnesium oxide $MgO(s) + 2H_2O(l) \rightleftharpoons Mg(OH)_2(aq)$
>
> **iii** Aluminium oxide: no reaction
>
> ### Part 2
>
> **c** As we go from left to right across the Periodic Table, the bonding in the chlorides changes from ionic to covalent. This is shown by the reactions with water. Ionic chlorides like NaCl will simply dissolve in water but covalent chlorides like $AlCl_3$ will hydrolyse to give the metal hydroxide and hydrogen chloride. Magnesium chloride has partial covalent character because of the high charge density on the Mg^{2+} ion.
>
> **d i** Sodium chloride $NaCl(s) + (aq) \rightarrow Na^+(aq) + Cl^-(aq)$
>
> **ii** Magnesium chloride: Two possible reactions depending on the results obtained.
> $MgCl_2(s) + (aq) \rightarrow Mg^{2+}(aq) + 2Cl^-(aq)$ Or
> $MgCl_2(s) + 2H_2O(l) \rightarrow Mg(OH)Cl(aq) + HCl(aq)$
>
> **iii** Aluminium chloride: $AlCl_3(s) + 3H_2O(l) \rightarrow Al(OH)_3(s) + 3HCl(g)$

Practical investigation 7.2:
Relative atomic mass of magnesium using a back-titration method

Introduction
In this investigation, learners will apply their existing knowledge regarding moles, titrations and relative atomic mass. A back-titration method uses the fact that the relative atomic mass of a metal means that we do not know how much acid will react with it. Excess standard acid is added to the metal and the excess acid is titrated against standard alkali. This will give the amount of acid remaining. As the starting amount of acid is known, the amount reacting with the metal can be calculated; from this the number of moles of metal can be calculated and from that its relative atomic mass.

Skills focus
The following skill areas are developed and practised (refer the skills grids at the front of this guide for codes):

MMO	Collection of data and observations (a), (b), (c), (d) and (e)
	Decisions relating to measurements of observations (a), (b) (c) and (d)
PDO	Recording data and observations (a), (c) and (e)
	Display of calculations and reasoning (a) and (b)
ACE	Data interpretation and sources of error (d) and (g)
	Drawing conclusions (c)

Duration
This practical will require half an hour to set up and one hour to complete.

Preparing for the investigation
- Ahead of this practical investigation, learners need to revisit relative atomic mass, moles and concentrations in solution. They will also need to refresh the procedures for making up solutions and titrations.

Equipment
Each learner or group will need:

- 50 cm^3 burette
- a small glass funnel for filling burette
- a larger glass funnel for preventing loss of acid spray
- a white tile for titration
- 25.00 cm^3 pipette
- pipette filler
- a pair of scissors and a ruler
- 50.00 cm^3 pipette or 25.00 cm^3 measuring cylinder.
- 250 cm^3 conical flask for the reaction vessel and for the titrations.
- dropper bottle filled with methyl orange indicator

Cambridge International AS & A Level Chemistry

Access to:
- standard 0.500 mol dm^{-3} hydrochloric acid
- standard 0.100 mol dm^{-3} sodium hydroxide solution
- magnesium ribbon
- steel wool
- top-pan balance reading to three decimal places

Safety considerations
- eye protection must be worn at all times.
- the sodium hydroxide is an irritant at the concentration provided
- when the magnesium reacts with the acid there is some acid spray formed but this is minimised by using the glass filter funnel
- when filling the burette with the standard alkali, care must be taken
- methyl orange is poisonous. If any is splashed onto skin it should be washed off immediately.

Carrying out the investigation
Ensure learners are aware of the two relevant balanced symbol equations and the information that can be gleaned from them. For example, make it clear that the number of moles of acid is equal to the number of moles of alkali. It should also be emphasised that the number of moles of magnesium present is equal to half the number of moles of hydrochloric acid that have reacted.

- It may be useful to go through the equations that are to be used in the practical calculations.

Common learner misconceptions
- Some learners get confused with the amount of acid that has reacted and the amount that remains. Once this is clarified then there should be no problems.
- Learners may also forget that once they have made up the solution to 250 cm^3 they are titrating one tenth of the total amount of acid remaining from the reaction. Therefore, they have to remember that they have to multiply by 10 in order to calculate the amount of acid that remains.

Sample results
Mass of magnesium ribbon = 0.160 g

See Table 7.3

Burette reading/cm^3	Rough	1	2	3
2nd	12.00	23.80	36.00	47.80
1st	0.00	12.00	23.80	36.00
Titre/cm^3	12.00	11.80	12.20	11.80

Table 7.3

Answers to the workbook questions (using the sample results)

a The volume of NaOH required to neutralise the HCl = 11.80 cm^3

b $n_{HCl} = n_{NaOH} = C \times V = 0.100 \times 11.80 \times 10^{-3} = 1.18 \times 10^{-3}$ mol

c n_{HCl}(remaining) = $10 \times 1.18 \times 10^{-3}$ (remember 25.00 cm^3 out of 250 cm^3 were titrated)

$= 1.18 \times 10^{-2}$ mol

d n_{HCl}(start) = $0.500 \times 50 \times 10^{-3}$ mol = 2.50×10^{-2} mol

n_{HCl}(that reacted) = n_{HCl}(start) − n_{HCl}(remaining) = $2.50 \times 10^{-2} - 1.18 \times 10^{-2} = 1.32 \times 10^{-2}$ mol

e Mg(s) + 2HCl(aq) ⟶ MgCl$_2$(aq) + H$_2$(g)

f $n_{Mg} = \frac{1}{2} n_{HCl} = \frac{1}{2} \times 1.32 \times 10^{-2}$ mol = 6.6×10^{-3} mol

g $A_r(Mg) = \frac{m}{n} = \frac{0.160}{6.6 \times 10^{-3}} = 24.2$ g mol^{-1}

h i Not all the acid spray was trapped in the flask

ii Some of the acid solution was not transferred from the reaction vessel to the volumetric flask

iii Errors in the measuring apparatus

iv Error in the estimation of the end-point of the titration

Practical investigation 7.3: Planning

Investigation into the separation of two metal ions in solution

Introduction
The aim is for the learners to plan and then carry out an investigation to separate the magnesium ions from a mixture of magnesium and barium ions in solution. They will then need to identify the magnesium ions.

Skills focus
The following skill areas are developed and practised (refer the skills grids at the front of this guide for codes):

MMO	Collection of data and observations (a), (b), (c) and (e)
	Decisions relating to measurements of observations (e) and (f)
ACE	Drawing conclusions (a), (c) and (d)
	Suggesting improvements (a) (b) and (c)

Duration
This practical will require half an hour to complete.

Preparing for the investigation
- This is a problem-solving exercise. Learners do not require any knowledge outside the concepts covered in the coursebook and the sheet on qualitative analysis in the syllabus specification.

Equipment
Each learner or group will need:
- two boiling tubes and one test tube
- test-tube rack
- filter funnel and filter paper
- two droppers
- a wash bottle filled with distilled water

Access to:
- a mixture of barium and magnesium ions: mixture should be composed of magnesium nitrate at $1.00 \, \text{mol dm}^{-3}$ and barium nitrate at $0.01 \, \text{mol dm}^{-3}$. Note: barium nitrate is not as soluble as magnesium nitrate and takes a lot longer to dissolve.
- $1.00 \, \text{mol dm}^{-3}$ sodium hydroxide solution
- $1.00 \, \text{mol dm}^{-3}$ sodium sulfate solution

Safety considerations
- The mixture contains barium ions, which are toxic. Any spillages must be wiped down with plenty of water and washed from the skin immediately.
- The sodium hydroxide solution is an irritant at the concentration provided.

Carrying out the investigation
Ask how learners could get the barium ions from the mixture.

Common learner misconceptions
- Learners can find interpreting the solubility tables problematic. The simple remedy is to look at the relative solubility (e.g. barium hydroxide is x750 more soluble than magnesium hydroxide in water and magnesium sulfate is nearly x200 000 more soluble than barium sulfate).

Sample results (learner method)
1. Step 1: Add a specified volume, for example $5 \, \text{cm}^3$, of the mixture to a boiling tube and to it add a few drops of the sodium sulfate solution.
2. Explanation: The barium ions are precipitated as barium sulfate. Magnesium sulfate is much more soluble and the magnesium ions remain in solution.
3. Step 2: Filter the mixture. The residue is barium sulfate and the filtrate contains magnesium ions.
4. Explanation: The insoluble solid barium sulfate is separated from the solution.
5. Take a small sample of the filtrate and to it add a few drops of sodium hydroxide solution. A white precipitate confirms the presence of magnesium ions.
6. See Table 7.4

Method step	Observations
1	A white precipitate is formed
2	The filtrate is colourless and the residue is a white solid
Identification	A white precipitate is formed

Table 7.4

Cambridge International AS & A Level Chemistry

> **Answers to the workbook questions (using the sample results)**
>
> **a** $Ba^{2+}(aq) + SO_4^{2-}(aq) \rightarrow BaSO_4(s)$
>
> $Mg^{2+}(aq) + 2OH^-(aq) \rightarrow Mg(OH)_2(s)$
>
> **b** Learners' assessment of their own method and explanation
>
> **c** Add sodium hydroxide solution to the mixture of ions. Filter the precipitate to separate the magnesium hydroxide, which will be the residue. Barium hydroxide is relatively soluble and will come through as the filtrate.

Practical investigation 7.4: Identification of three metal compounds using qualitative analysis

Introduction

In this investigation, learners are asked to identify three unknown compounds of a Group 2 metal using a number of qualitative tests. They should identify both the metal ion and the negative ions present in each compound. Each compound contains three elements. The metal ion is magnesium.

Skills focus

The following skill areas are developed and practised (refer the skills grids at the front of this guide for codes):

MMO   Collection of data and observations (a), (b), (c) and (e)
Decisions relating to measurements of observations (f)

ACE   Drawing conclusions (c) and (d)

Duration

This practical requires one hour to complete.

Preparing for the investigation

- Ahead of this practical investigation, learners should review thermal decomposition of Group 2 nitrates.
- Part 1 requires a test for carbon dioxide. The learners may not remember how to test for this gas. Ask them to review the Skills chapter if that is the case.

Equipment

Each learner or group will need:

- five test tubes, two boiling tubes and a test-tube rack
- four droppers
- filter funnel and three filter papers
- Bunsen burner and heat-resistant pad
- wooden splint
- wash bottle filled with distilled water

Access to:

- A (magnesium carbonate), B (magnesium hydroxide) and C (magnesium nitrate) for testing
- 2.00 mol dm^{-3} hydrochloric acid
- 1.00 mol dm^{-3} nitric acid
- lime water
- Universal Indicator (U.I.) solution
- a fume cupboard

Safety considerations

- Eye protection must be worn at all times during the investigation.
- During the heating of any solids make sure learners do not inhale any gases evolved and place the test tube in the fume cupboard.
- The U.I. is dissolved in ethanol and is therefore flammable.
- The lime water is an irritant.
- The acids are irritants at the concentrations provided.

Carrying out the investigation

- Heating compound C (magnesium nitrate) will lead to the formation of nitrogen dioxide. The solid needs very little heating but the nitrogen dioxide is toxic. As soon as it is observed, the learners should also test for oxygen. Once the tests and observations are made then the test tube should be placed in the fume cupboard.

  Ask learners to suggest another magnesium compound and suitable tests for it (e.g. magnesium sulfate or magnesium chloride).

Chapter 7: The properties of metals

Sample results

Part 1: Investigating compound A

See Table 7.5

Method step	Observations	Conclusions
1	Effervescence. The gas evolved gave a white precipitate with limewater (turned limewater cloudy)	A is a carbonate – contains the CO_3^{2-} ion
2	There is an effervescence	
3	The filtrate is colourless	
4	White precipitate formed	This suggests that the ion present is probably Mg^{2+}

Table 7.5

Part 2: Investigating compound B

See Table 7.6

Method step	Observations	Conclusions
1	Not applicable	Not applicable
2	B is insoluble in water	B is not magnesium sulfate or magnesium nitrate
3	The U.I. turns blue	B is a weak alkali

Table 7.6

Part 3: Investigating compound C

See Table 7.7

Method step	Observations	Conclusions
1	C dissolves in water	C is a soluble magnesium compound. It could be magnesium sulfate or magnesium nitrate.
2	U.I. solution turns green	C is a neutral salt of magnesium
3		
4	C gives off a dark brown gas. A glowing splint relights in the gas	The brown gas is nitrogen dioxide and oxygen is also formed because it is the gas that relights the glowing splint. This shows that C is a nitrate.

Table 7.7

Answers to the workbook questions (using the sample results)

Part 1

a **i** $2HCl(aq) + MgCO_3 \rightarrow MgCl_2(aq) + H_2O(l) + CO_2(g)$

Or $2H^+(aq) + CO_3^{2-}(s) \rightarrow H_2O(l) + CO_2(g)$

ii $CO_2(g) + Ca(OH)_2(aq) \rightarrow CaCO_3(s) + H_2O(l)$ (the equation for the test for carbon dioxide)

iii $2HNO_3(aq) + MgCO_3 \rightarrow Mg(NO_3)_2(aq) + H_2O(l) + CO_2(g)$

Or $2H^+(aq) + CO_3^{2-}(s) \rightarrow H_2O(l) + CO_2(g)$

iv $Mg^{2+}(aq) + 2OH^-(aq) \rightarrow Mg(OH)_2(s)$

b Compound A is magnesium carbonate ($MgCO_3$)

Part 2

c $Mg(OH)_2(s) + (aq) \rightarrow Mg^{2+}(aq) + 2OH^-(aq)$ (equilibrium lies well over to the left)

d B is magnesium hydroxide $Mg(OH)_2$

Part 3

e $2Mg(NO_3)_2(s) \rightarrow 2MgO(s) + 4NO_2(g) + O_2(g)$

f C is magnesium nitrate

Chapter 8:
The properties of non-metals

Chapter outline

This chapter relates to Chapter 12: Group 17 and Chapter 13: Nitrogen and sulfur in the coursebook.

In this chapter learners will complete practical investigations on:

- 8.1 Formula of hydrated sodium thiosulfate crystals
- 8.2 Preparation and properties of the hydrogen halides
- 8.3 Reaction of bromine with sulfite ions (sulfate (IV))
- 8.4 Identification of unknowns containing halide ions

Practical investigation 8.1:
Formula of hydrated sodium thiosulfate crystals

Introduction

Thiosulfate–iodine titrations are a key category of titration that your learners should be able to carry out. This investigation will also allow them to master the concept of linked reactions. The investigation is quite challenging. In particular, learners will need to use their judgement as to when to add the starch indicator. The reaction end-point is straightforward to identify. The calculations required will give learners a chance to review their knowledge of moles in solution and the relationship between mass, numbers of moles and formula mass.

Skills focus

The following skill areas are developed and practised (refer the skills grids at the front of this guide for codes):

MMO	Collection of data and observations (a), (b), (c), (d) and (e)
	Decisions relating to measurements of observations (a), (b) and (d)
PDO	Recording data and observations (a) (b), (c), (d) and (e)
	Display of calculations and reasoning (a) and (b)
	Data layout (a)
ACE	Drawing conclusions (c)

Duration

This practical will require two hours to complete. The first hour is required for the preparation of the learners with regards to the linking of the two reactions involved and the preparation of the two solutions, i.e. the copper(II) sulfate and the sodium thiosulfate. The second hour will be taken up with the titrations and the data analysis.

Preparing for the investigation

- This investigation is probably best done in groups, but it can be completed by single learners. One of the limiting factors is that each experiment requires $2 \times 250 \, cm^3$ volumetric flasks.

- This is probably the first time learners will come across linked reactions involving iodimetry. Before they do this investigation, learners should be informed about linked reactions and why they are so useful.

- Please ensure learners also know how to gauge when to add the starch indicator.

- The copper (II) sulfate should be of ANALAR quality.

Equipment

Each learner or group will need:

- $150 \, cm^3$ conical flask
- $2 \times 250 \, cm^3$ volumetric flasks
- 1% starch indicator and dropper
- wash bottle filled with distilled water

Chapter 8: The properties of non-metals

- burette stand
- 25.0 cm³ pipette
- white tile
- 250 cm³ beaker and 100 cm³ beaker
- stirring rod
- small dropper
- small filter funnel for burette and larger one for volumetric flask
- 10 cm³ measuring cylinder

Access to:
- copper(II) sulfate ($CuSO_4.5H_2O$) solution
- 0.100 mol dm^{-3} hydrochloric acid
- sodium thiosulfate solution
- 0.500 mol dm^{-3} potassium iodide solution
- a two or three place top-pan balance

Safety considerations

- The copper(II) sulfate solution is harmful and is an environmental hazard.
- Eye protection should be worn at all times in the practical.

Carrying out the investigation

- Because the iodide is in excess, a measuring cylinder is an adequate way of measuring out the required volume of potassium iodide solution.
- The main challenge is when to add the starch indicator and in particular what is meant by a pale straw colour. It is recommended that this is demonstrated.
- The precipitate of copper(I) iodide makes the identification of the straw colour and the end-point slightly more difficult to identify.
- The purity of sodium thiosulfate can be an issue but results are usually good.

- The final calculation does require the use of algebra since the sodium thiosulfate has a relative formula mass of 158.4 + 18x. Some of the less mathematically capable students may need to be given some priming for such calculations.
- Use of equivalent numbers of moles will help to overcome the problem of the linked reaction.
- Practising problems that require the calculation of the number of moles of water of crystallisation may be helpful.
- Encourage learners to write their results in pencil unless they are particularly confident.

Common learner misconceptions

- Some learners think that the thiosulphate ion reacts directly with the copper(II) ion.
- Some learners think that a pale straw colour indicates the end-point, which is incorrect.

Sample results

Part 1: Preparation of solutions

Mass of $CuSO_4.5H_2O$ = 3.13 g

Mass of sodium thiosulfate crystals = 6.20 g

Part 2: Titration

See Table 8.1

Burette reading/ cm³	Rough titration	1st	2nd	3rd
2nd	13.00	25.60	37.95	32.90
1st	0.00	13.00	25.60	20.30
Titre/cm³	13.00	12.60	12.35	12.60

Table 8.1

Answers to the workbook questions (using the sample results)

a The concordant titres are 12.60 cm³ and 12.60 cm³ The average of these values = 12.60 cm³

b 2 mol of Cu^{2+}(aq) ≡ 1 mol of I_2(aq)

c 1 mol of I_2(aq) ≡ 2 mol of $S_2O_3^{2-}$(aq)

d 2 mol of Cu^{2+}(aq) ≡ 2 mol of $S_2O_3^{2-}$(aq); therefore 1 mol of Cu^{2+}(aq) ≡ 1 mol of $S_2O_3^{2-}$(aq);

e i $M_r(CuSO_4.5H_2O) = 249.6$ g mol⁻¹

   Number of moles of copper(II) ion in 250 cm³
   $= \dfrac{3.13}{249.6} = 0.01254$ mol;
   Concentration of copper(II) ion $= \dfrac{n}{V} = \dfrac{0.01254}{0.25}$
   $= 0.0502$ mol dm⁻³

   Number of moles of copper(II) ion in 25.00 cm³ = $C \times V = 0.0502 \times 25 \times 10^{-3} = 0.001255$ mol

   ii Number of moles of thiosulfate ion = number of moles of copper(II) ion = 0.001255 mol

   The volume of thiosulfate solution = 12.60 cm³;

   Concentration of thiosulfate solution
   $= \dfrac{n}{V} = \dfrac{0.001255}{12.6 \times 10^{-3}} = 0.0996$ mol dm⁻³

   Therefore number of moles of thiosulfate ion in 250 cm³ = 0.0998 × 0.250 = 0.0249 mol

   iii Mass of sodium thiosulfate dissolved in 250 cm³ of solution = 6.20 g

   Therefore 0.0249 mol of sodium thiosulfate crystals weigh 6.20 g

   Therefore 1 mol of sodium thiosulfate crystals weigh $\dfrac{6.20}{0.0249}$ g mol⁻¹ = 248.5 g mol⁻¹

   Total formula mass of crystals = mass of $Na_2S_2O_3$ + mass of x moles of H_2O

   = 158.4 + 18x; Therefore 158.4 + 18x = 248.5;
   18x = 248.5 − 158.4 = 90.1; x = $\dfrac{90.1}{18} = 5$

Practical investigation 8.2:
Preparation and properties of the hydrogen halides

Introduction
Learners will prepare hydrogen chloride, hydrogen bromide and hydrogen iodide and then investigate their chemical reactions.

Skills focus
The following skill areas are developed and practised (refer the skills grids at the front of this guide for codes):

MMO   Collection of data and observations (a), (b), (c) and (e)
      Decisions relating to measurements of observations (a), (c) and (d)

ACE   Drawing conclusions (c) and (d)

Duration
This practical will take one and a half hours to complete. The first half an hour is needed to go through the health and safety aspects. The second hour is needed for the preparation and testing of the hydrogen halides.

Preparing for the investigation
The apparatus used for preparing and collecting hydrogen halides needs to be absolutely dry. The addition of phosphorus(V) oxide is one way of making sure that the phosphoric acid is as dry as possible.

If corks are used to stopper the test tubes, they should be gently dried beforehand to make sure that they do not contain any water.

Equipment
Each learner or group will need:

- approximately 15 **dry** test tubes and a test-tube rack.
- stoppers or corks for test tubes

Chapter 8: The properties of non-metals

- three dry boiling tubes
- three right-angled glass delivery tubes
- plastic gloves
- Bunsen burner, heat-resistant pad and straight tongs
- small spatula
- short length of nichrome wire
- thin glass stirring rod
- paper towels
- retort stand, boss and clamp
- 250 cm^3 beaker or small trough

Access to:
- concentrated phosphoric acid
- nitric acid
- silver nitrate solution
- phosphorous (V) oxide / phosphorous pentoxide
- Universal Indicator (U.I.) in dropper bottles
- solid potassium chloride, solid potassium bromide, and solid potassium iodide
- concentrated ammonia solution in a small dropper bottle
- a top-pan balance

Safety considerations

- Both the concentrated phosphoric acid and phosphorous (V) oxide are corrosive and should be treated with great care.
- Because of the corrosive nature of the reactants and products, safety spectacles do not give sufficient protection; safety goggles must be worn at all times.
- The hydrogen halide fumes themselves are harmful and should not be inhaled. When the learners are trying to find out whether or not the test tubes are full of gas they can blow over the mouths of the test tubes for a brief moment. No one should be on the other side facing the person doing the exhalation. Once the misty fumes of the acid are visible; the test tube should be stoppered and replaced by the next one.
- The ammonia is a highly pungent gas and it is best to test its reaction with the hydrogen halides in a fume cupboard. The fumes of the ammonium halide that is formed are also harmful.
- At the end of the experiment the boiling tubes should be placed in the fume cupboard where they can be washed up. When water is added to the contents a lot of heat is evolved and it is probably best that the teacher or technician carry out the clearing up of the boiling tubes.

Carrying out the investigation

- Learners will need some help in ascertaining when the test tubes are full of hydrogen halide gas. Blowing across the mouth of the test tube very briefly is sufficient to give misty fumes if the tube is full of gas.
- The solubility of the hydrogen halides is very well illustrated by the rise of the water in the test tubes. If the teacher so wishes, this can be emphasised by a demonstration of the solubility in sodium hydroxide solution. The wearing of plastic gloves would be essential in such a demonstration.
- The reactions of the hydrogen halides with ammonia tend to form quite dense fumes of ammonium halide and therefore if a fume cupboard is available then it must be used. The fume cupboard must also be the final resting place of any apparatus that has been used for the collection and preparation of the hydrogen halides.
- The formation of the hydrogen halides gets harder from HCl to HBr and then to HI.
- It can take some time before the hydrogen halide gas comes off in sufficient quantities and the first test tubes may not be full of gas.
- The reaction with water is a very good indicator as to whether the tube is full of gas; if the water rises right up the tube it means that the gas is displaced by it.
- 🔬 Before they start, learners should be asked to read through the practical and discuss what they think the objectives of the practical are. How does it relate to what they are learning or have learned about Group 17?
- The words 'trends and patterns' could be used. What are the trends in the properties of the hydrogen halides? What are the patterns?
- 💡 Hydrogen fluoride is a weak acid; hydroiodic acid is a very strong acid. Ask learners to explain this.

Common learner misconceptions

- The solubility in water is not due to dissolving. This is a chemical change – the formation of H_3O^+ and halide ions; dissolving is a physical change.

Cambridge International AS & A Level Chemistry

Sample results

See Table 8.2

Hydrogen halide gas	Observations			
	Test tube 1: upside down test tube in water	Test tube 2: reaction with ammonia	Test tube 3: action of heat	Test tube 4: silver nitrate
HCl	Water rises far up the test tube. U.I. turns red	Dense white fumes are formed	No change	White precipitate which dissolves in dilute ammonia
HBr	Water rises far up the test tube. U.I. turns red	Dense white fumes are formed	No change	Pale-cream precipitate which does not dissolve in dilute ammonia but does in conc. ammonia
HI	Water rises far up the test tube. U.I. turns red	Dense white fumes are formed	Brown–purple colour seen	Yellow precipitate that is insoluble in dilute and concentrated precipitate

Table 8.2

Answers to the workbook questions (using the sample results)

a $H_3PO_4 + KX \longrightarrow HX + KH_2PO_4$

b The water reacts with the hydrogen halides. It is an acid–base reaction because the HX donates a proton, and is therefore an acid, to a water molecule, which acts as a base. The $H_3O^+(aq)$ ion (hydroxonium ion) is formed along with the halide ion.

$HX(g) + H_2O(l) \longrightarrow H_3O^+(aq) + X^-(aq)$

c The white fumes are solid ammonium halides. This is another acid–base reaction. The HX donates a proton (acid) to the ammonia (base) to give the ammonium halide.

$HX(g) + NH_3(g) \longrightarrow NH_4X(s)$

d Only the hydrogen iodide dissociates to form the halogen (iodine) and hydrogen. The bond enthalpies of HCl and HBr are too high for the heat generated by the hot wire to have any effect.

e All three hydrogen halides react with water to give the halide ions. These ions can be detected using silver nitrate solution to give precipitates of the silver halide.

i $Ag^+(aq) + Cl^-(aq) \longrightarrow AgCl(s)$

ii $Ag^+(aq) + Br^-(aq) \longrightarrow AgBr(s)$

iii $Ag^+(aq) + I^-(aq) \longrightarrow AgI(s)$

Practical investigation 8.3:
Reaction of bromine with sulfite ions (sulfate (IV))

Introduction

Learners investigate the reaction between aqueous bromine and sulfite ions (sulfate (IV))

Skills focus

The following skill areas are developed and practised (see the skills grids at the front of this guide for codes):

MMO   Collection of data and observations (b), (c) and (e) Decisions relating to measurements of observations (a), (b) (c) and (f)

PDO   Recording data and observations (e)

ACE   Drawing conclusions (a), (c) and (d)

Duration

This practical requires one hour to complete.

Equipment

Each learner or group will need:

- boiling tube
- three test tubes and a test-tube rack
- two droppers
- plastic gloves

Chapter 8: The properties of non-metals

- small spatula
- wash bottle filled with distilled water
- 10 cm³ measuring cylinder

Access to:

- bromine water
- sodium sulfite solid
- 0.100 mol dm⁻³ barium chloride solution
- 2.00 mol dm⁻³ hydrochloric acid
- 2.00 mol dm⁻³ nitric acid
- 0.100 mol dm⁻³ silver nitrate solution
- 2.00 mol dm⁻³ ammonia solution
- concentrated ammonia solution

Safety considerations

- The bromine water is harmful and should not be held near the nose or face at any time.
- All of the acids are irritants at the concentrations provided.
- The silver nitrate is an irritant at the concentration provided.
- Solutions of barium ions are toxic. The barium chloride should be handled with care and, if possible, plastic gloves should be worn.

Carrying out the investigation

- The control test for sulfite can sometimes give a residual faint white precipitate after reaction with barium chloride solution and subsequent addition of hydrochloric acid. This can be explained away by the oxidation of the sulfite to sulfate by atmospheric oxygen.
- If the sulfite is quite old then this problem may mask the oxidation of sulfite by the bromine.

Common learner misconceptions

- Learners can think of oxidation in the simplest terms, i.e. the gain of oxygen and this is one such situation. The SO_3^{2-} ion gains one oxygen atom to form the SO_4^{2-} ion. They should be encouraged to think of it in terms of either an increase in oxidation state of the sulfur, from +4 to +6 or by loss of electrons as shown below.

$$H_2O(l) + SO_3^{2-}(aq) \longrightarrow 2H^+(aq) + SO_4^{2-}(aq) + 2e^-$$

Sample results

See Table 8.3

Reaction	Observations
Step 2 Bromine water + sodium sulfite solution	The bromine is decolourised
Step 4: Portion I Addition of barium chloride solution followed by dilute hydrochloric acid	A white precipitate is formed that does not dissolve when hydrochloric acid is added
Step 5: Portion II Addition of silver nitrate and nitric acid followed by dilute ammonia solution, then conc. ammonia solution	A pale-cream precipitate is formed. This did not dissolve in dilute ammonia solution but did dissolve in concentrated ammonia solution.

Table 8.3

Answers to the workbook questions (using the sample results)

a The bromine decolourised.

b The SO_4^{2-} (sulfate(VI)) ion is formed.
   $Ba^{2+}(aq) + SO_4^{2-}(aq) \longrightarrow BaSO_4(s)$

c The Br⁻ ion. The pale cream precipitate with silver nitrate solution was partially soluble in ammonia solution (did not dissolve in dilute aqueous ammonia solution but did dissolve in concentrated aqueous ammonia).

   $Ag^+(aq) + Br^-(aq) \longrightarrow AgBr(s)$

d $Br_2(aq) + SO_3^{2-}(aq) + H_2O(l) \longrightarrow 2Br^-(aq) + SO_4^{2-}(aq) + 2H^+(aq)$
   (+2 / −2)

e The bromine is reduced because the oxidation state of each Br atom decreases from 0 to −1. The sulfur is oxidised because its oxidation number increases from +4 to +6.

Practical investigation 8.4: Identification of unknowns containing halide ions

Introduction
Three unknown compounds (X, Y and Z) need to be provided so that the learners can carry out specific tests and use the results to identify the ions present. All three compounds should contain the same cation.

Skills focus
The following skill areas are developed and practised (see the skills grids at the front of this guide for codes):

MMO	Collection of data and observations (b), (c) and (e)
	Decisions relating to measurements of observations (c), (e) and (f)
PDO	Recording data and observations (e)
ACE	Drawing conclusions (c) and (d)

Duration
- This practical requires one hour to complete.

Preparing for the investigation
- This investigation requires learners to be aware of qualitative analysis tests.

Equipment
Each learner or group will need:
- five test tubes plus a test-tube rack
- Bunsen burner and heat-resistant pad
- three droppers
- Universal Indicator paper
- wash bottle of distilled water
- stoppers for test tubes
- small spatula

Access to:
- unknown solids X, Y and Z
- X = ammonium iodide
- Y = ammonium chloride
- Z = ammonium bromide
- $2.00\,mol\,dm^{-3}$ sodium hydroxide solution
- $2.00\,mol\,dm^{-3}$ ammonia solution (dilute ammonia solution)
- Concentrated ammonia solution
- $0.100\,mol\,dm^{-3}$ silver nitrate solution
- $2.00\,mol\,dm^{-3}$ nitric acid
- cyclohexane
- chlorine water (this is prepared by the reaction of concentrated hydrochloric acid with solid potassium manganate(VII) and passing the chlorine through water)
- fume cupboard

Additional notes and advice
- If any of the unknown solids are not available, simply mix ammonium nitrate with the potassium halide. This will give the desired results in the test.

Safety considerations
- Sodium hydroxide solution is corrosive at the concentration used in the investigation.
- The nitric acid is an irritant.
- The dilute ammonia solution is an irritant.
- The concentrated ammonia solution is harmful and should not be taken out of the fume cupboard.
- Cyclohexane vapour is harmful. Do not dispose of this down the sink. Use the reagent bottle available and decant the upper layer into this bottle.
- The silver nitrate is an irritant
- The chlorine water is a saturated solution and will give chlorine, which is toxic. Avoid inhalation. As with the concentrated ammonia, it should be kept in the fume cupboard or in a stoppered boiling tube.

Carrying out the investigation
- Chloride and iodide are easy to distinguish. It is the bromide ion which is the most difficult to distinguish because of its insolubility in dilute ammonia but solubility in concentrated ammonia. If the learners find this aspect difficult then an alternative approach would be to try the dilute ammonia first and then take a fresh solution of the unknown and investigate the solubility in concentrated ammonia solution.

Chapter 8: The properties of non-metals

- The learners are asked to add ammonia solution of the appropriate concentration. For Cl⁻ this is dilute ammonia solution; for I⁻ this is concentrated ammonia solution and for Br⁻ this is dilute ammonia followed by concentrated ammonia solution. These additions should be informed by the results obtained with the displacement reactions.

If help is required to understand the test for the ammonium ion, it can be viewed as the reverse reaction for the reaction of ammonia with water, i.e.
$NH_3(aq) + H_2O(l) \rightarrow NH_4^+(aq) + OH^-(aq)$

Sample results

See Table 8.4

Method step	Observations of unknowns		
	X	Y	Z
1	There is a strong pungent smell and the gas turns the moist U.I. paper blue	Same as for X	Same as for X
2	The solution turns brown/a purple solid forms. When the cyclohexane is added the brown colour of the aqueous layer fades and the cyclohexane layer turns purple.	The solution turns yellow in colour. When the cyclohexane is added the yellow colour fades in the aqueous layer and there is no change in the cyclohexane layer.	The solution turns yellow/orange in colour. When the cyclohexane is added the yellow/orange colour fades in the aqueous layer and the cyclohexane layer turns orange.
3	A yellow precipitate forms. The precipitate is insoluble in concentrated ammonia solution.	A white precipitate forms. The precipitate dissolves in dilute ammonia solution.	A pale-cream precipitate forms. The precipitate is insoluble in dilute ammonia solution but dissolves in concentrated ammonia solution.
Amount and strength of ammonia solution added	Amount: a few (3–5) drops Strength: concentrated ammonia	Amount: 2–5 drops of Strength: dilute ammonia	Amounts: 2–3 drops of dilute then 2–5 drops of concentrated. Strength: dilute **then** concentrated ammonia

Table 8.4

Answers to the workbook questions (using the sample results)

a Ammonium ion (NH_4^+)

b The gas is ammonia indicated by the smell. Ammonia is an alkaline gas.
Reaction equation: $NH_4^+ + OH^- \rightarrow NH_3 + H_2O$

c Iodide I⁻

d i Iodide is displaced by the chlorine to form iodine. This is confirmed by the purple colour in cyclohexane. $2I^- + Cl_2 \rightarrow 2Cl^- + I_2$

  ii The presence of I⁻ is confirmed by the yellow precipitate of silver iodide which is insoluble in ammonia.

  $Ag^+(aq) + I^-(aq) \rightarrow AgI(s)$

e Cl⁻ chloride ion

f i Chloride cannot be displaced by chlorine so there is no reaction

  ii The presence of Cl⁻ is confirmed by the white precipitate of silver chloride which is soluble in ammonia (dissolves in dilute ammonia).

  $Ag^+(aq) + Cl^-(aq) \rightarrow AgCl(s)$

g Bromide Br⁻

h i Bromide is displaced by the chlorine to form bromine. This is confirmed by the orange colour in cyclohexane.

  $2Br^- + Cl_2 \rightarrow 2Cl^- + Br_2$

  ii The presence of Br⁻ is confirmed by the pale-cream precipitate of silver bromide which is partially soluble in ammonia because it did not dissolve in dilute ammonia but did dissolve in concentrated ammonia solution.

  $Ag^+(aq) + Br^-(aq) \rightarrow AgBr(s)$

Hydrocarbons and halogenoalkanes

Chapter outline

This chapter relates to Chapter 15: Hydrocarbons and Chapter 16: Halogenoalkanes in the coursebook.

In this chapter learners will complete practical investigations on:

- 9.1 Cracking of hydrocarbons
- 9.2 How halogenoalkane structure affects the rate of hydrolysis

Practical investigation 9.1:
Cracking of hydrocarbons

Introduction

In this investigation the learners will investigate the thermal decomposition of hydrocarbons and evaluate the economic importance of the process.

Skills focus

The following skill areas are developed and practised (see the skill grids at the front of this guide for codes):

MMO	Collection of data and observations (a), (b), (c) and (e)
	Decisions relating to measurements of observations (a), (b) (c), and (d)
PDO	Recording data and observations (a), (b), (d) and (e)
	Data layout (a)
ACE	Drawing conclusions (c) and (d)

Duration

The testing of the paraffin oil should take no more than 30 minutes to complete. It may be necessary to use the rest of the lesson (assuming each lesson is at least one hour) to give the learners a thorough instruction on how to carry out the cracking. This would make it more likely that they will complete the practical work in the following lesson without hurrying too much.

Preparing for the investigation

- Any test tubes used for the cracking investigation cannot be used for any other experiment. Therefore, if cracking has been done before, it is suggested that you re-use the same test tubes.

- You should be able to purchase the paraffin oil from a hardware store or pharmacy.

- A Bunsen valve is required on the end of the delivery tube. This is used to decrease the possibility of suck-back and learners should be shown what to do if suck-back occurs.

- A demonstration of the cracking apparatus set-up is a good idea.

Equipment

Each learner or group will need:

- Bunsen burner, tripod and gauze
- heat-resistant mat
- heat-resistant test tube (e.g. Pyrex®)
- Delivery tube plus stopper. The delivery tube needs to have a Bunsen valve at the end
- small trough
- several (at least five) test tubes plus stoppers
- wooden splint
- dropper
- spatula
- retort stand, boss and clamp
- plastic (vinyl) gloves for handling the ceramic wool

Access to:

- paraffin oil
- ceramic wool
- bromine water
- broken pot

Chapter 9: Hydrocarbons and halogenoalkanes

Safety considerations

- The learners will be heating a test tube to very high temperatures and therefore must be reminded to allow several minutes for the apparatus to cool down before handling it.
- Bromine water is harmful and must be handled with care.
- There is a danger of suck-back. While the Bunsen valve is designed to minimise this danger, it could still happen and they must be aware of this (see Carrying out the investigation section).
- The products of cracking can cause irritation to airways. Vapours produced must be gently 'wafted' towards the nose rather than inhaled.
- Ceramic wool can cause skin irritation and plastic gloves should be worn when using the material.
- The products of the reaction between the bromine water and the gaseous products of cracking can be harmful and should be disposed of by washing down the sink using large amounts of water.

Carrying out the investigation

- When learners are familiar with and confident about tackling suck-back, this investigation is straightforward.
- For those less confident, this investigation enables learners to improve their procedures when heating substances.

  🔧 Learners can make sense of their findings by referring to balanced equations used to represent cracking (e.g. $C_{12}H_{26} \rightarrow C_8H_{18} + 2C_2H_4$).
  $\quad\quad\quad\quad\quad\quad\quad\quad\quad$ A $\quad\quad$ B

  Substance A would give a positive result when tested with a lit splint, as would substance B. Substance B is the compound responsible for the reaction with bromine water.

- Remind learners that a popping sound indicates the production of hydrogen gas in the reaction (e.g. $C_8H_{18} \rightarrow 4C_2H_4 + H_2$).

  🔧 Ask learners to write as many equations as they can to show the cracking of, for example, C_8H_{18}.

- Ask them to draw the apparatus that could be used to collect a volatile liquid product of cracking as well as the gaseous product.

Sample results

Learners will prepare their own results table, anticipated to be similar to Table 9.1

Substance tested	Tests	Observations
Paraffin oil	Smell	No smell
	Lit splint	The lighted splint has to be held very close to the paraffin oil and takes a very long time to light. Burns slowly
	Bromine water	No reaction. The orange colour of the bromine water is concentrated into the paraffin oil.
Products of cracking	Smell	Distinct smell of petrol
	Lit splint	The gas lights very easily and burns with a blue flame. Sometimes there is a popping sound when it lights.
	Bromine water	The bromine water is decolourised immediately after shaking

Table 9.1

Answers to workbook questions (using the sample results)

a i Paraffin oil does not light easily or burn quickly so it is not a good fuel.

ii Paraffin oil is not unsaturated and therefore cannot undergo addition polymerisation.

b i The product of cracking would make a good fuel. It has a low boiling point, so is easily vaporised. It ignites easily and burns quickly.

ii The product of cracking decolourises bromine, so it is unsaturated. Therefore, it can undergo addition polymerisation.

c The paraffin oil is a poor fuel and cannot be used to make addition polymers. Therefore, the paraffin oil is not very useful and has very little economic value. The products of cracking are good fuels and can be used to make addition polymers. Therefore, by using cracking, the economic value of the paraffin oil is greatly increased.

Practical investigation 9.2:
How halogenoalkane structure affects the rate of hydrolysis

Introduction
Learners will investigate how the structure of a halogenalkane affects the rate of its hydrolysis. They will complete this by investigating the relative rates of hydrolysis of primary, secondary and tertiary chloroalkanes.

Skills focus
The following skill areas are developed and practised (see the skill grids at the front of this guide for codes):

MMO	Collection of data and observations (a), (b), (c) and (e)
	Decisions relating to measurements of observations (c) and (f)
PDO	Recording data and observations (a)
	Display of calculations and reasoning (a)
	Data layout (a) and (b)
ACE	Data interpretation and sources of error (a), (b), (d) and (f)
	Drawing conclusions (a), (c) and (d)

Duration
This practical should take no more than one hour to complete.

Preparing for the investigation
- Chloro-derivatives are used, but this investigation works equally well (if not better) with the bromo-derivatives.
- This investigation is a good test of learners' ability to organise themselves. It may be a good idea to ask learners to discuss in groups how they should complete this practical before they start.
- If there are insufficient numbers of stopwatches then one stopwatch can be used and the mixing of the solutions can be at staggered intervals such as 1 minute.

Equipment
Each learner or group will need:
- 250 cm³ beaker
- −10 to 110 °C thermometer
- six test tubes and three stoppers
- four dropping pipettes
- wooden splint
- permanent marker pen
- three stopwatches
- two 10 cm³ measuring cylinders
- test-tube rack
- glass or plastic stirring rod

Access to:
- ethanol
- 0.100 mol dm⁻³ silver nitrate solution
- 1-chlorobutane, 2-chlorobutane and 2-chloro-2-methylpropane
- boiling water (ideally a kettle)

Safety considerations
- Because the investigation requires flammable compounds (e.g. ethanol and the halogenoalkanes), learners should not have any naked flames near their reaction containers.
- If a kettle is unavailable or too many groups need hot water at the same time, water can be heated using Bunsen burners. Please ensure that these are turned off before any dispensing of flammable liquids takes place.
- The silver nitrate should be considered as harmful and an irritant and therefore should be handled with care.

Carrying out the investigation
- The main problem is organising the mixing of silver nitrate solution with the ethanol and halogenoalkane. If the learners are aware of what has to be done then they can organise themselves in the best way they can in order to be efficient.
- The reaction with the primary halogenoalkane is slow and the formation of the precipitate takes a long time. For this reason, it is sometimes difficult to gauge when the cross on the wooden splint has been obscured, as the process is so gradual.
- The reaction with the tertiary halogenoalkane is very quick and the instant when the cross is obscured can easily pass without being realised. Both this point and the previous one are part of the evaluation.

Chapter 9: Hydrocarbons and halogenoalkanes

⚙ Some learners have difficulty in understanding why there should be ethanol and silver nitrate solutions. If necessary you can demonstrate the fact that a miscible liquid is required for a reaction to occur using an oil-water mixture.

⚙ Expect learners to work out the relative rates without any assistance.

Common learner misconceptions

- Some learners have difficulty in understanding why $\frac{1}{Time}$ is used to measure rate.

Sample results

Learners will prepare their own results table, anticipated to be similar to Table 9.2

Compound	Time taken to obscure cross/s	Rate of reaction (1/time) s^{-1}	Relative rate
1-chlorobutane	609	0.00164	1
2-chlorobutane	85	0.0118	7.2
2-chloro-2-methylpropane	12	0.0833	50.8

Table 9.2

Answers to the workbook questions (using sample results)

Part 1

a i In the reaction between the water and the chloroalkanes, chloride ions are formed and the silver ions react with these to form precipitates of silver chloride. The speed at which these precipitates are formed show how quickly the reaction proceeds.

$RCl(l) + H_2O(l) \longrightarrow ROH(l) + H^+(aq) + Cl^-(aq)$
$Ag^+(aq) + Cl^-(aq) \longrightarrow AgCl(s)$

ii The chloroalkanes are immiscible with aqueous solutions. Therefore, if they are added to the silver nitrate solution, two layers will form and reaction will only occur where the two layers meet. The ethanol acts as a solvent for both the silver nitrate solution and the chloroalkanes allowing the reactants to mix.

Part 2

b Time taken for the precipitate to obscure the cross on the wooden splint.

c See Table 9.2. Learners use the value for 1/time taken for the precipitate to obscure the cross on the wooden splint. This is because the time taken is a measure of how slow the reaction is, therefore 1/time is a measure of how quick it is.

d See Table 9.2

e The tertiary chloroalkane undergoes a unimolecular (S_N1) reaction where the rate-determining step is where the chloroalkane dissociates to give the chloride ion and a carbonium ion.

$(CH_3)_3CCl \longrightarrow (CH_3)_3C^+ + Cl^-$ slow
$(CH_3)_3C^+ + OH^- \longrightarrow (CH_3)_3COH$ fast

This means that the reaction rate depends on how quickly the first reaction happens. The primary chloroalkane undergoes a bimolecular (S_N2) reaction, which depends on the OH^- ion and chloroalkane molecules coming together so that as the water molecule approaches, the chlorine atom leaves as a chloride ion. This is much slower than the unimolecular reaction because it depends on the molecule and the ion colliding in just the right way on the carbon atom attached to the chlorine atom.

$RCl + OH^- \longrightarrow HO^-\text{-----}R\text{-----}Cl$ slow
$HO^-\text{-----}R\text{-----}Cl \longrightarrow ROH + Cl^-$ fast

f i The reaction for the tertiary chloroalkane is so quick that the incident at which the cross is obscured can be easily missed.

ii The reaction for the primary chloroalkane is so slow and so gradual that it is difficult to say when exactly the cross is obscured.

iii If the reactions are started at different times the temperature of the reaction vessels may be slightly different and this will affect the rate of reaction.

g See Table 9.3

Control variable	Why the variable is kept constant
Number of carbons in compounds	To make sure it is the structure that is tested and not the number of carbons
Concentration of silver nitrate solution	So that the rate that the precipitate forms is dependent on the structure of the halogenoalkane and not the concentration of reactants. If the concentration of a reactant varies so does the rate.
Temperature	Temperature will affect the rate of formation of the precipitate
Number of drops of halogenoalkane	Equal concentrations of halogenoalkanes

Table 9.3

Chapter 10:
Organic compounds containing oxygen

Chapter outline

This chapter relates to Chapter 17: Alcohols, esters and carboxylic acids and Chapter 18: Carbonyl compounds in the coursebook.

In this chapter, learners will complete an investigation on:

- 10.1 Identifying four unknown organic compounds

Practical Investigation 10.1:
Identifying four unknown organic compounds

Introduction

The investigation has three parts, consisting of different tests that will enable the identification of four organic compounds that should be provided:

P: propan-2-one (CH_3COCH_3)

Q: propanoic acid (CH_3CH_2COOH)

R: propanal (CH_3CH_2CHO)

S: propan-2-ol ($CH_3CH(OH)CH_3$)

Skills focus

The following skill areas are developed and practised (see the skill grids at the front of this guide for codes):

MMO	Collection of data and observations (a), (b), (c) and (e)
	Decisions relating to measurements of observations (a), (c) and (d)
PDO	Recording data and observations (a), (b) and (c)
	Data layout (a)
ACE	Drawing conclusions (a), (c) and (d)

Duration

The practical investigation will take approximately three hours to complete in its entirety.

A suggested sequence for completion is as follows:

- Parts 1 and 2a will take about one hour to demonstrate and complete.
- Parts 2b and 2c will take a further one hour to complete.
- Part 3 and its required planning will take a final one hour to complete.

Preparing for the Investigation

- Learners should observe Part 1 of the investigation, then complete Part 2 on compounds Q and S and Part 3 on compounds P and R.
- You may decide to demonstrate some of the other procedures ahead of the learners completing them.
- If access to any of the suggested test compounds is limited or you have insufficient amounts, then you can use other compounds that will give positive results in the same tests. For example, propan-2-ol could be replaced by ethanol as it will give the identical reaction in the test with phosphorus pentachloride; it undergoes esterification with glacial ethanoic acid and it will give a positive result in the iodoform test.
- The learners should be aware of the theory accompanying the reactions of the functional groups before they do the practical.

Part 1: Test for hydroxyl groups using phosphorus pentachloride

This part of the investigation is an observed demonstration

Equipment

For the practical demonstration you will need:

- four unknown solids: P, Q, R and S.
- four dry test tubes
- test-tube rack
- spatula
- Universal Indicator (U.I.) paper
- phosphorus pentachloride
- fume cupboard

Safety considerations

- The demonstration must take place in a fume cupboard.
- Ensure learners wear eye protection and tie long hair back and that they stand at least two metres away from the fume cupboard during the demonstration.

Carrying out the investigations

- The phosphorus pentachloride will react with any moisture present and give misty fumes of hydrogen chloride. Therefore, it is vital that the compounds tested are dry. Anhydrous calcium chloride can be used to dry the compounds.

Sample results

See Table 10.1

Unknown compound	Observations
P	There is no reaction
Q	Misty fumes are formed which turn moist Universal Indicator paper red. An exothermic reaction takes place.
R	There is no reaction
S	Misty fumes are formed which turn moist Universal Indicator paper red. An exothermic reaction takes place.

Table 10.1

Answers to the workbook questions (using sample results)

a Compounds Q and S contain hydroxyl groups. Compounds P and R do not.

b Phosphorus pentachloride reacts with -OH groups to give hydrogen chloride. Hydrogen chloride is an acidic gas which will turn Universal Indicator paper red.

$ROH + PCl_5 \longrightarrow RCl + POCl_3 + HCl$

Part 2: Investigating the compounds that do contain a hydroxyl group

Equipment

Each learner or group will need:

- six test tubes
- test-tube rack
- Bunsen burner and heat-resistant pad
- wooden splint
- spatula
- two evaporating basins
- graduated droppers
- 250 cm^3 glass beaker

Access to:

- samples of the unknown compounds that tested positive for a hydroxyl group in Part 1
- sodium hydrogen carbonate
- sodium carbonate solution
- 2.00 mol dm^{-3} sodium hydroxide solution
- limewater solution
- wash bottle filled with distilled water
- concentrated sulfuric acid in a dropper bottle
- porcelain dish or a white tile
- glacial ethanoic acid
- hot water (kettle)
- iodine solution

Cambridge International AS & A Level Chemistry

Safety considerations

- Eye protection must be worn at all times and tie long hair back.
- The limewater is an alkali and should be treated as corrosive.
- Sodium hydroxide solution is corrosive.
- The organic compounds are flammable and must be kept away from naked flames.
- The organic compounds must also be regarded as being harmful. If possible plastic gloves should be worn to minimise contact.
- The concentrated sulfuric acid is corrosive. Always add concentrated sulfuric acid to water **never** the other way around. If you get any acid on your skin, wash off immediately using large amounts of cold water.
- Hot water should be provided by a kettle.
- The product of the reaction in Part 2c is strongly irritating to eyes. As soon as you have made your observations, wash the reaction mixture down the sink with plenty of water.
- The iodine solution will stain the skin so handle with care.

Carrying out this part of the investigation

- The Tollens' reagent test needs to be carried out using a new test tube if possible. If these are not available then older test tubes should be thoroughly washed in distilled water, rinsed in propan-2-one and then dried. If carried out carefully, this test will give a very nice silver mirror on the inside surface of the tube.
- If a precipitate is not forming in the 2,4-DNPH test for the carbonyl compounds, adding one or two drops of sulfuric acid usually starts the process.
- 👤 Learners can be given more groups of compounds and asked to formulate a process whereby these can be identified in a similar way.
- The equation for the oxidation of aldehydes using Tollens' reagent is rather simplistic and does not really show what is happening. An alternative approach would be to explain to more able learners how the silver ions are reduced by accepting electrons from the aldehyde.

    $CH_3CH_2CHO + H_2O \rightarrow CH_3CH_2COOH + 2H^+ + 2e^-$

    $Ag^+ + e^- \rightarrow Ag$

    Overall $CH_3CH_2CHO + H_2O + 2Ag^+ \rightarrow CH_3CH_2COOH + 2H^+ + 2Ag$

- Learners could also be asked to draw a key that will help them organise their thoughts regarding the different reactions of the functional groups, for example, draw up their own scheme by testing for the C=O group. If the answer to 'does it react with 2,4-DNPH?' is 'yes', they can use Tollens' reagent. If the answer is 'no' they can test for the –COOH group or for an alcohol.

Part 2a: Test for carboxylic acid group

Sample results

See Table 10.2

Unknown compound	Observations
Q	When the sodium hydrogen carbonate is added to the aqueous solution of Q there is effervescence and the gas formed turns the limewater cloudy
S	There is no reaction

Table 10.2

> ### Answers to the workbook questions (using sample results)
>
> **c** Compound Q contains the carboxylic acid group (-COOH).
>
> **d** Compound Q reacts with sodium hydrogen carbonate to give carbon dioxide gas. This is a typical reaction of the aqueous solution of a carboxylic acid.
>
> **e** The compound is propanoic acid (CH_3CH_2COOH). The equation for the reaction is as follows:
>
> $CH_3CH_2COOH(aq) + NaHCO_3(s) \rightarrow CH_3CH_2COO^-Na^+ (aq) + H_2O(l) + CO_2(g)$

Part 2b: Test for alcohol: the R–OH group

Sample results

- When the reaction mixture is added to the sodium carbonate solution there is fizzing and the mixture gives off a sweet fruity smell.
- There is an oily liquid covering the sodium carbonate solution. This shows that an ester has been formed.

Chapter 10: Organic compounds containing oxygen

> **Answers to the workbook questions (using sample results)**
>
> **f** An ester
>
> **g** The compound must be an alcohol because alcohols react with glacial ethanoic acid to form esters.
>
> There are two alcohols containing three carbons in their molecules – propan-1-ol ($CH_3CH_2CH_2OH$) and propan-2-ol ($CH_3CH(OH)CH_3$).

Part 2c: Iodoform reaction: test for $CH_3CH(OH)-$ group or the CH_3CO- group

Sample results

- A yellow precipitate forms slowly

> **Answers to workbook questions (using sample results)**
>
> **h** The compound is propan-2-ol because if a compound gives the iodoform reaction then it must contain the $CH_3CH(OH)-$ group and propan-2-ol contains this group but propan-1-ol does not.

Part 3: Identifying the compounds that do <u>not</u> contain the hydroxyl group

Sample results

Equipment

Each learner or group will need:

- two test tubes
- three graduated droppers
- permanent marker pen

Access to:

- 2,4-di-nitrophenylhydrazine (2,4-DNPH or Brady's reagent) solution in methanol

Safety considerations

- 2,4-DNPH and the methanol it is dissolved in are toxic. Methanol is also flammable.
- Eye protection must be worn at all times and tie long hair back.

Part 3a: Reaction with 2,4-di-nitrophenylhydrazine

Sample results

- P = yellow–orange precipitate
- R = yellow–orange precipitate

> **Answers to the workbook questions (using sample results)**
>
> **i** Both P and R are carbonyl compounds. They are either aldehydes or ketones.

Part 3b: Reaction with Tollens' reagent

Sample results

- P = there is no reaction
- R = a black precipitate/silver mirror is formed

> **Answers to the workbook questions (using sample results)**
>
> **j** Both compounds contain three carbons. P must be a ketone because it does not give a silver mirror with Tollens' reagent. P must be propan-2-one (CH_3COCH_3). R is an aldehyde because it gives a silver mirror with Tollens' reagent. R must be propanal (CH_3CH_2CHO).
>
> **k** P = propan-2-one (CH_3COCH_3)
>
> Q = propanoic acid (CH_3CH_2COOH)
>
> R = propanal (CH_3CH_2CHO)
>
> S = propan-2-ol ($CH_3CH(OH)CH_3$)

More about enthalpy changes

Chapter outline

This chapter relates to Chapter 4: Chemical bonding, Chapter 6: Enthalpy changes and Chapter 19: Lattice energy in the coursebook.

In this chapter, learners will complete investigations on:

- 11.1 Enthalpy change of vaporisation of water
- 11.2 Enthalpy change of solution of chlorides
- 11.3 Planning investigation into thermal decomposition of iron(II) ethanedioate
- 11.4 Planning investigation into thermal decomposition of metal carbonates
- 11.5 Data analysis investigation into enthalpy change of mixing

Practical investigation 11.1:
Enthalpy change of vaporisation of water

Extension investigation
Skills focus

The following skill areas are developed and practised (see the skills grids at the front of this guide for codes):

PI	Control experiments and identification of variables a,c
COI	Methods used (a, b, c, d) Carrying out the experiment (b)
HI	Collecting and displaying data (b, c) Manipulating data (b, c)
DA	Identifying and using calculations (b)
CP	Drawing conclusions (e)
EI	Identifying problems with the procedure (a, e) Making a judgement on the conclusions (a)

Duration

The practical work will take 30 minutes. The analysis and evaluation questions will take 30 minutes.

Preparing for the investigation

- Learners should have some experience of enthalpy changes and the relationship $E = mc\Delta T$ from previous studies.

- Before starting the experiment, learners should know how to measure the temperature of a liquid accurately and understand how to use a Bunsen burner to get a non-luminous flame.

Equipment

Each learner or group will need:

- clamp stand, two clamps and two bosses
- Bunsen burner
- 500 cm^3 conical flask. It is important that the flask is a hard glass silicate type which will not break on direct heating. A large beaker could be used as an alternative.
- 100 cm^3 or 250 cm^3 measuring cylinder

- 0–100 °C long-stemmed thermometer. This should be approximately 15–20 cm in length
- cork or rubber bung with hole bored to fit thermometer
- stopclock or stopwatch
- glass rod (preferably with a rubber top) for stirring or wire stirring loop

Access to:
- distilled water
- means of lighting the Bunsen burner

Alternative equipment and method

- electric kettle
- stopclock or stopwatch
- 500 cm^3 measuring cylinder

This method has the disadvantage that there is no opportunity to plot a graph.

The procedure is:

1. Record the power, in watts, of the kettle. This should be on the base plate of the kettle.
2. Put 400 cm^3 of distilled water in the kettle and switch on, leaving the lid off.
3. Start the stopclock when the water begins to boil.
4. Let the water boil for 4 min.
5. Leave the kettle to cool then measure the volume of water remaining in the kettle.

The enthalpy change of vaporisation is given by:

$$\frac{\text{electrical energy supplied / kJ}}{\text{amount of water boiled away / mol}}$$

The electrical energy supplied can be calculated using the relationship:

$$\text{kilowatts} = \frac{\text{kilojoules}}{\text{seconds}}$$

Safety considerations

- Learners should be aware that steam causes burns.
- Learners should be encouraged to take care when reading the thermometer and stirring the water.
- If a rubber end for the glass stirring rod is not available, learners could hold the end of the rod with a small piece of paper towel.

Carrying out the investigation

- Learners determine the enthalpy change of vaporisation by measuring the energy required to heat up and then boil away a particular mass of water.
- The experiment depends on the energy from the Bunsen burner flame being supplied at a constant rate. Therefore, it is very important to keep the size of the flame constant and to protect it from draughts.
- The height of the flame should be arranged so that it does not touch the bottom of the flask directly or only the very tip touches the flask. If a larger area of flame touches the flask, there is a possibility that the flask will break during the heating.
- Ensure that learners have a Bunsen flame which is entirely non-luminous, otherwise deposits of soot will build up on the bottom of the flask and prevent heat getting to the water at a constant rate. If available, small butane gas cylinders often produce a better flame. A beaker could be used in place of the flask and may give better results.
- A flask is used here because of the evaluation question **i**. Ensure that the thermometer will not move during the investigation. This can be done using a cork or rubber bung with a hole through which the thermometer fits.
- An alternative is to clamp the thermometer using several paper towels folded around the top but not obscuring the scale.
- Only gentle stirring of the water is needed, just before the temperature readings are taken. Stirring should not be vigorous, as there is a possibility of breaking the thermometer: A simple up and down movement is all that is required. A wire stirring loop is preferable in this respect.
- ⚙ Some learners may need help in setting up the apparatus to the correct height and adjusting the Bunsen flame to the correct height.
- You might consider setting up the apparatus for them in advance.
- Learners may also be given hints for questions **c** (1 cm^3 water has mass of 1 g), **i** (think about the shape of the flask) and **m** (give a hint about the heating effect of an electric current).
- ⚙ Learners could be given information about electrical heating; for example, watts = joules per second and you could suggest in more detail how to carry out this experiment.
- Learners who finish early could also help struggling learners plot a suitable graph or undertake the calculations.

Cambridge International AS & A Level Chemistry

Common learner misconceptions

- The experiment is straightforward and the only misconception may arise from incorrect use of the Bunsen burner.

Sample results

Table 11.1 gives an idea of the results learners may obtain in the investigation.

Temperature values beyond the boiling point have not been included.

Volume of water remaining at the end of the experiment is 154.0 cm^3

Time / minutes	0	$\frac{1}{2}$	1	$1\frac{1}{2}$	2	$2\frac{1}{2}$	3	$3\frac{1}{2}$	4	$4\frac{1}{2}$	5	$5\frac{1}{2}$	6
Temp/°C	18	23	31	40	44	54	61	68	74	84	89	94	100

Table 11.1

Answers to the workbook questions (using the sample results)

a Graph should be linear cutting the x-axis at about 5.7 minutes. The points nearer the boiling point may be below the line because of greater heat losses.

b Gradient of graph: $\dfrac{100-18}{5.7} = 14.39\,°C\,min^{-1}$

c energy min^{-1} = $\dfrac{200(g) \times 4.18 \times 82}{5.7}$ (°C)

$= 12027\,J\,min^{-1}$

(assuming that the mass of 1 cm^3 of water is 1 g)

d by simple proportion: = 12027 × 10 = 120270 J

e moles = $\dfrac{mass\,water}{M_r\,water} = \dfrac{200-154}{18} = 2.555\,mol$

f $\dfrac{120270}{2.555} = 47072\,J\,mol^{-1}$ (approximately 47 kJ mol^{-1})

g Heat will be transferred directly to it by conduction and the reading will be too high.

h The mass of 1 cm^3 of water is 1 g; the heating is being applied at constant rate.

i Value is higher because some of the steam condenses on the cooler neck of the flask. Therefore, the mass of water boiled off is less than expected so the number of moles of water in part **f** is lower than expected.

j Weigh the water instead of measuring the volume since the measuring cylinders can be accurate to only about 1 cm^3 (or less if using a 250 cm^3 cylinder).

k Bunsen flame is not steady: Reduce air draughts by putting a few boards arranged vertically around the Bunsen burner. Bunsen may produce carbon on bottom of flask, which can reduce energy transfer: Make sure that the tip of the Bunsen burner flame is not too close to the flask. Lack of steady stirring: Leads to temperature 'hot spots'.

l Ethanol is flammable so should not be used near naked flames.

m Use electrical heating / use a heating element.

Practical investigation 11.2:
Enthalpy change of solution of chlorides

Skills focus

The following skill areas are developed and practised (see the skills grids at the front of this guide for codes):

COI	Methods used (c)
	Carrying out the experiment (a, b, c, d)
HI	Collecting and displaying data (b, c)
	Manipulating data (b, c)
DA	Identifying trends and patterns (a)
CP	Drawing conclusions (a, e)
	Making predictions (a, b)
EI	Identifying problems with the procedure (a)
	Making a judgement on the conclusions (a)

Duration

The practical work will take 20 minutes; the analysis and evaluation questions will take 25 minutes.

Chapter 11: More about enthalpy changes

Preparing for the investigation

- Learners should have some experience of endothermic and exothermic enthalpy changes from their previous studies. In this investigation, learners measure the enthalpy change of solution when some ionic solids are dissolved in water.

- Before starting the experiment, learners should be aware of how to weigh compounds accurately and of the importance of precise measurements

Equipment

Each learner or group will need:

- 250 cm^3 expanded polystyrene cup and beaker (see diagram in workbook)
- lid with hole for thermometer to fit the polystyrene cup
- 20 cm^3 (or 10 cm^3) measuring cylinder
- −10 to 100 °C thermometer (preferably with 0.1 °C graduations)

Access to:

- distilled water
- balance to weigh to at least one decimal place
- weighing boats
- anhydrous lithium chloride in stoppered container with spatula
- anhydrous sodium chloride in stoppered container with spatula
- anhydrous potassium chloride in stoppered container with spatula
- anhydrous magnesium chloride in stoppered container with spatula
- anhydrous calcium chloride in stoppered container with spatula

Additional notes and advice

- Calcium chloride will absorb moisture from the air and dissolve (it is deliquescent). It should be dried using a desiccator. Bottles of this salt should not be left open. The salts should be absolutely dry.

- In order to avoid contamination of salts, you might consider putting a small amount of each salt into a small beaker or boiling tube together with a spatula, each labelled with the name of the specific salt. The beakers or tubes should be stoppered to avoid moisture being absorbed.

Safety considerations

- Anhydrous calcium chloride is an irritant.
- The other chlorides are low hazard.

Carrying out the investigation

- In this investigation, learners measure the enthalpy change of solution when some ionic solids are dissolved in water.

- There are few problems in carrying out this experiment.

- Some of the changes in temperature are small, so careful reading of the thermometer is essential.

- Learners should be advised to stir the solution with the thermometer very carefully to avoid breaking the thermometer by pressing it on the bottom of the polystyrene cup.

- Advise learners not to swirl the cup in case any crystals get trapped on the side and do not dissolve. In order to get reasonably good results, the dissolving should not take more than 2–3 min, otherwise heat losses are too great.

- Some learners may need help in extrapolating the graph to determine a corrected temperature change in data analysis question **b**.

- They may also need to be given a hint in question **g** about the relative temperatures of the surroundings and the solution in this endothermic reaction.

- Learners may not immediately realise in question **j** that the solutions formed are fairly concentrated and this does not quite fit with the definition, which refers to very dilute (infinitely dilute) solutions.

- If more able learners finish early, they could be asked to explain the relationship between the enthalpy change of solution and the position of the chlorides in Group I, in terms of differences in lattice energy and hydration energy.

Common learner misconceptions

- This is a simple investigation that should pose few problems to learners.

Cambridge International AS & A Level Chemistry

Sample results

Tables 11.2 and 11.3 give an idea of results that learners may obtain in the investigation using about 0.04 moles of salt per 20 cm³ water.

Maximum temperature changes on adding the salts to water.

Chloride	Maximum temperature change
LiCl	+12.5 °C
NaCl	−0.9 °C
KCl	−7.0 °C
$MgCl_2$	+40.2 °C
$CaCl_2$	+28.5 °C

Table 11.2

Data for the energy change for potassium chloride.

Time / min	0	½	1	1½	2	2½	3	3½	4	4½	5	5½	6
Temperature / °C	19.7	19.8	19.9	19.8	19.8	14.2	13.6	13.5	13.6	14.0	13.9	14.2	14.5

Table 11.3

Answers to the workbook questions (using the sample results)

a See Tables 11.2 and 11.3.

b Please refer to Figure 11.1

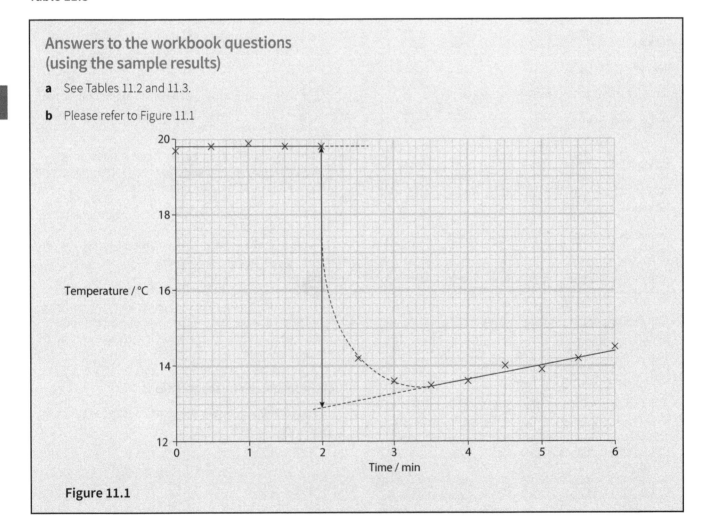

Figure 11.1

> **c** Energy (J) = 20 (g) × 4.18 × 7 (°C) = 585. J
>
> This assumes that the specific thermal capacity of the solution is the same as the specific thermal capacity of water.
>
> **d** energy / mol^{-1} = $\frac{585.2}{\text{moles KCl}}$ = $\frac{585.2}{0.04}$ = +14 630 J mol^{-1}
>
> (14.6 kJ mol^{-1})
>
> This value is lower than the actual value of + 16 kJ mol^{-1} because of heat losses.
>
> **e** As you go down the Group, the enthalpy change of solution changes from exothermic to endothermic and then becomes even more endothermic. The chlorides tend to absorb heat more as the size of the cation increases.
>
> **f** The Group II chlorides dissolve with a greater (exothermic) energy change than the corresponding Group I halides in the same Period. This can be related to a more highly charged cation and a greater number of chloride ions.
>
> **g** The surroundings are at a higher temperature than the solution so heat is being gained by the solution from the surroundings all the time after mixing. The graphical method attempts to compensate for this heat gain by extrapolating to the point at which the solid was added.
>
> **h** A burette or graduated pipette could have been used to dispense the water instead of the less accurate measuring cylinder.
>
> **i** The sample of solid could be placed in a clean dry test tube in some water in a second polystyrene cup for a few minutes. The water should be from the same container as used for the main experiment since it will be at the same temperature.
>
> **j** The definition suggests that the enthalpy change refers to **very** dilute solutions. The solutions used here are fairly concentrated: 0.04 mol / 20 cm^3 is a 2.0 mol dm^{-3} solution. Further dilution of this solution will absorb (or release) more energy but we would need a very accurate thermometer to record this!

Practical investigation 11.3: Planning
Investigation into thermal decomposition of iron(II) ethanedioate

Skills focus
The following skill areas are developed and practised (see the skills grids at the front of this guide for codes):

PI	Selecting information (c)
	Defining the problem under investigation (a, c, d)
	Considering hazards (a)
COI	Methods used (a, b, d)
	Carrying out the experiment (b)
HI	Collecting and displaying data (a, b)
DA	Identifying and using calculations (a, b)
CP	Drawing conclusions (b, e)
EI	Making a judgement on the conclusions (a)

Duration
The planning will take 30 minutes; the analysis and evaluation questions will take 30 minutes.

Preparing for the investigation
- Learners should have some experience of setting up apparatus for measuring gas volumes and doing calculation involving the molar gas volume. It would be helpful beforehand to give a lesson on drawing chemical apparatus.
- Learners could suggest an equipment list and do some of the planning for prep. They could also research how acidic gases can be absorbed.
- Some learners who have not had much experience with practical work have a limited idea of the capacity of common laboratory apparatus such as gas syringes, or have difficulties relating the amounts of reagents used to the volume of gas collected. A demonstration of the capacity and limitations of various pieces of laboratory apparatus before the investigation is carried out can be useful.

Equipment
Suitable apparatus that could be chosen by each learner:
- gas syringe (or 100 cm^3 measuring cylinder and trough of water)
- hard glass test tube or boiling tube
- delivery tube with bung to fit the test tube / boiling tube

- Bunsen burner
- clamp stands and clamps
- bottle / U-tube or other container to hold the absorbent for carbon dioxide
- connecting tubes to join test tube / carbon dioxide absorber and gas syringe
- solution to absorb carbon dioxide, for example, concentrated aqueous sodium hydroxide/ damp calcium oxide.
- balance to weigh to nearest 0.01 g

Method

- A suitable apparatus set-up (shown in Figure 11.2) and method chosen by each learner might be as follows.

1. Weigh the test tube to the nearest 0.01 g.
2. Put an appropriate mass of iron(II) ethanedioate in the test tube and reweigh to nearest 0.01 g.
3. Connect the apparatus as shown with the gas syringe plunger fully in or the measuring cylinder full of water (or record the initial volumes).
4. Heat the iron(II) ethanedioate gently at first and then more strongly.
5. When no more gas has been collected, record its volume.
6. When heating has stopped, immediately disconnect the test tube from the rest of the apparatus to avoid suck-back.
7. Allow the test tube to cool then reweigh it to the nearest 0.01 g.

Safety considerations

- This investigation is a planning exercise and should not be done practically since carbon monoxide is produced in the reaction and although contained in the apparatus could be present in sufficient quantity to cause severe headaches or worse.
- Learners should answer the safety consideration questions given the information in the workbook: The experiment should be a closed system with no leaks so that no carbon monoxide escapes. The experiment should be carried out in a fume cupboard to minimise any poisonous gas getting into the laboratory. The iron(II) ethanedioate should be weighed out using gloves and making sure that no dust is raised. Concentrated sodium hydroxide is corrosive so care is needed dispensing it. Iron(II) oxalate is low hazard and although learners are not told of its hazard rating, they could make a comment to be careful of substances they are not sure of. They could also look up information from CLEAPS A level cards for prep.

Carrying out the investigation

- The learners are asked to plan an experiment to show that the molar ratio of iron(II) oxide and carbon monoxide agrees with the equation which is shown in the workbook. The learners should suggest absorbing the carbon dioxide in an alkali, measuring the volume of carbon dioxide gas produced as well as weighing the mass of iron(II) oxide, which is the other product of the decomposition.
- When drawing chemical apparatus, many candidates leave gaps or draw lines across tubes where there should be a free flow of gas in the mistaken idea that as long as the basic apparatus is present, it does not matter how it is drawn or if it is labelled or not.

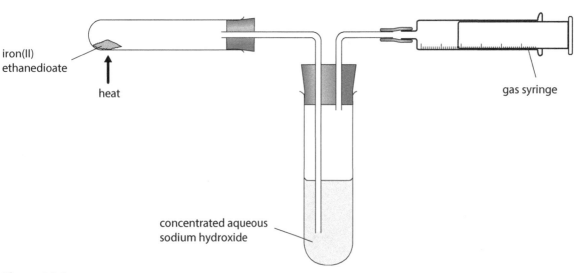

Figure 11.2

Chapter 11: More about enthalpy changes

- 🛠 Some learners may need help in drawing diagrams of the apparatus. The use of a ruler or a suitable stencil may help here.

- Others may have difficulty in organising the sequence of steps to be followed in the method section of the experiment. They could be encouraged to plan the sequence on a separate sheet of paper first and then rearrange as necessary.

- Some learners may need some help in parts **b** and **c**, where the appropriate volume should be large enough to minimise errors but not so large as to be greater than or equal to the volume of the syringe or measuring cylinder.

- Some learners may need help in answering part **e**, where the mass and hence the number of moles of iron(II) oxide should be calculated. Reference to the equation and the answer to part **c** should confirm that the number of moles of carbon monoxide and iron(II) oxide should be equal.

- 💡 Learners could be asked to suggest other thermal decomposition experiments where two different gases are formed, for example, the decomposition of Group II nitrates or unreactive nitrates such as lead(II) nitrate. They could also be asked to find out why a similar method to the one they have used in this investigation will not work for the decomposition of ammonium nitrate or ammonium carbonate.

- More able learners who finish early could also help struggling learners undertake the calculations when their own calculations have been checked.

Sample results

In this investigation, the learners are asked to select suitable values for the volume of gas to be collected. The rest of the results follow on from this. The volume suggested is 80 cm³ for a 100 cm³ gas syringe.

Answers to the workbook questions (using the sample results)

a 100 cm³ (if gas syringe or measuring cylinder)

b 80 cm³

c mol CO = $\dfrac{80}{24000}$ = 3.3×10^{-3} mol

d 1 mol CO produced from one mol $Fe(COO)_2$.

Molar mass of $Fe(COO)_2$ = 144.0

So mass = $3.3 \times 10^{-3} \times 144.0$ = 0.475 g

e Weigh the mass of iron(II) oxide formed at the end of the experiment (after cooling).
Then calculate the number of moles of iron(II) oxide. If this is equal to 3.3×10^{-3} mol, it is in the correct molar ratio.

f Keep heating until the volume of gas remains constant and until there is a constant mass of solid in the test tube (after cooling).

g Check the tubing/connection for leaks.

h It should be effective because CO_2 is 100 times more soluble in aqueous solutions than CO and therefore should be absorbed by the sodium hydroxide leaving the CO to go to the gas syringe. Percentage error caused by any CO dissolving is very small. It may not be effective if a balance reading to one decimal place is used because weighing errors are then considerable.

Practical investigation 11.4: Planning
Investigation into thermal decomposition of metal carbonates

Skills focus
The following skill areas are developed and practised (see the skills grids at the front of this guide for codes):

PI Defining the problem under investigation (a, c, d)
    Control experiments and identification of variables (c)
    Considering hazards (a)

COI Methods used (a, b, c, d)
    Carrying out the experiment (a, b, c, d)

HI Collecting and displaying data (a)
    Manipulating data (b)

CP Drawing conclusions (a, b)

EI Identifying problems with the procedure (a)
    Making a judgement on the conclusions (a)

Duration
The planning will take 30 minutes (or 90 minutes if practical work is included); the analysis and evaluation questions will take 20 minutes.

Preparing for the investigation
- Learners should have some experience of setting up apparatus for heating and should be familiar with the test for carbon dioxide using limewater (aqueous calcium hydroxide). They should also be able to distinguish between qualitative and quantitative data.
- Learners could suggest an equipment list and do the safety analysis for prep.

Equipment
Each learner or group will need:
- five hard glass test tubes or boiling tubes
- a test tube to for the aqueous calcium hydroxide
- right-angled delivery tube with bung to fit the test tube / boiling tube
- Bunsen burner
- clamp stands and clamps
- stopclock or stopwatch

Access to:
- Balance to weigh to nearest 0.1 g and weighing boats and spatulas

Method
A suitable apparatus set-up (see Figure 11.3) and method chosen by a learner could be as follows.

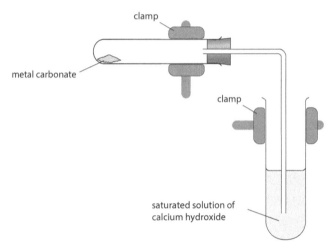

Figure 11.3

1. Weigh out a specified number of moles copper(II) carbonate.
2. Put the copper(II) carbonate in the test tube and place delivery tube and bung in the mouth of the test tube.
3. Place measured volume of aqueous calcium hydroxide in another test tube.
4. Connect the apparatus up as shown in the diagram.
5. Heat the copper(II) carbonate strongly using a roaring Bunsen flame and immediately start the stopclock.
6. Time how long it takes until the first sign of cloudiness in the aqueous calcium hydroxide and then how long it takes until the cloudiness disappears.
7. When heating has stopped, immediately disconnect the test tube from the rest of the apparatus to avoid suck-back.
8. Repeat the experiment using the same molar quantities of the other carbonates, the same volume of aqueous calcium hydroxide and the same amount of heating (same size and intensity of flame).

Chapter 11: More about enthalpy changes

Safety considerations

- Learners should answer the safety consideration questions given the information in the workbook. For example: gloves should be worn when dispensing lead(II) carbonate and iron(II) carbonate.

Carrying out the investigation

Some learners may need help in drawing diagrams of the apparatus. The use of a ruler or a suitable stencil may help here.

Others may have difficulty in organising the sequence of steps to be followed in the method section of the experiment. They could be encouraged to plan the sequence on a separate sheet of paper first and then rearrange as necessary.

Some learners may need some help with the mole calculation in part **a**. In part **c**, some learners may need to be given a hint that qualitative data indicates that specific numbers should not be used.

Some learners may need help in answering part **g**, where (amongst other things) the effectiveness of the procedure requires some reference to impurities in the copper(II) carbonate and oxidation of iron(II) carbonate.

Learners could be asked to look at coursebooks or use the internet to find out about the thermal decomposition of compounds such as sulfates and hydroxides to see if there are any patterns in common with those of the carbonates or nitrates.

Common learner misconceptions

- When drawing chemical apparatus, many candidates leave gaps or draw lines across tubes where there should be a free flow of gas in the mistaken idea that as long as the basic apparatus is present, it does not matter how it is drawn or if it is labelled or not.

Sample results

In this investigation, learners may get results which are wide ranging depending on the strength of the Bunsen burner flame. Typical result ranges using a fairly low Bunsen burner flame are: Copper(II) carbonate: ½–3 minutes; Iron(II) carbonate: 1–4 minutes; Lead(II) carbonate: ½–2 minutes; Magnesium(II) carbonate: 4–8 minutes; Sodium carbonate: no decomposition

---

Answers to the workbook questions (using the sample results)

a mol of $Ca(OH)_2$ in $100\,cm^3$ solution $= \dfrac{1.53 \times 10^{-2}}{10} = 1.53 \times 10^{-3}\,mol$

Molar mass of $Ca(OH)_2 = 74.1$ so mass $= 1.53 \times 10^{-3} \times 74.1 = 0.113\,g$

b Height and strength of Bunsen burner flame: the energy should be applied at a consistent rate for each experiment. The volume of aqueous calcium hydroxide: a smaller volume will cause a precipitate sooner because the carbon dioxide will be more concentrated.

c Please refer to Table 11.4

Carbonate	Time taken / minutes (first run)	Time taken / minutes (second run)	Average time taken /minutes
copper(II) carbonate			
iron(II) carbonate			
lead(II) carbonate			
magnesium carbonate			
sodium carbonate			

Table 11.4

d   Please refer to Table 11.5

Carbonate	Ease of decomposition
copper(II) carbonate	easy
iron(II) carbonate	fairly easy
lead(II) carbonate	easy
magnesium carbonate	difficult
sodium carbonate	no decomposition

Table 11.5

e   The time taken for the white precipitate to be observed is subjective. There is no standard depth of cloudiness to make a comparison. The time taken for the cloudiness to disappear gives a better indication, but this is also subjective.

f   To avoid suck-back due to the contraction of the volume of air in the tube when heating stops. The cold water sucked back will crack the tube.

g   The method used is unlikely to be effective because of the difficulty in determining when the cloudiness has been formed in the aqueous calcium hydroxide. It is difficult to keep the rate of heating the same and it takes time for the tube to heat up to a constant temperature. The copper(II) carbonate contains copper(II) hydroxide impurity, which does not produce carbon dioxide when it decomposes. The number of moles of the solid decomposed is less than it would be if pure carbonate was used. The iron(II) carbonate may oxidise during the reaction so we may be comparing a mixture of the iron(II) and iron(III) compounds.

Practical investigation 11.5: Data analysis
Investigation into enthalpy change of mixing

Skills focus

The following skill areas are developed and practised (see the skills grids at the front of this guide for codes):

PI Selecting information (b, c)
    Control experiments and identification of variables (a, b)
    Considering hazards (a)

COI Methods used (b, c)

HI Collecting and displaying data (b)

DA Identifying trends and patterns (a, b, c)

EI Identifying problems with the procedure (a, b)
    Identifying problems with the data (a, b, c)
    Making a judgement on the conclusions (b)

Duration

The data analysis and evaluation questions will take 35 minutes.

Preparing for the investigation

- Learners should revise intermolecular forces (Chapter 4 in the coursebook).

Safety considerations

- The experiment should be conducted in an enclosed area such as a tank with glove holes for manipulating the apparatus. If the fume cupboard is suggested, it should not be switched on until necessary to reduce the evaporation of the volatile liquids. It would be inadvisable to carry out the experiment in the open laboratory unless well ventilated.

- Gloves should be worn when dispensing the liquids. The liquids should be dispensed in the fume cupboard.

- Methanol is highly flammable, so no naked flames should be present.

Method

This is given in the introduction to the investigation within the workbook. It is not advisable that the experiment be carried out in a school laboratory.

Carrying out the investigation

- Learners are asked to analyse data about the enthalpy change when two liquids, trichloromethane and methanol, are mixed in different proportions. They are then asked to draw conclusions about the results.

Chapter 11: More about enthalpy changes

- In many cases when a liquid mixture is formed, one type of intermolecular force is replaced by another, which may be stronger or weaker than the attractive forces in the individual liquids.

  🛈 Some learners may need help in calculating the masses in part **e** and the mole calculation in part **f**.

- Others may need help interpreting the graph in part **g** in terms of the differences in the intermolecular forces in the pure liquids and the mixture.

  🛈 Learners who finish early can suggest how the enthalpy change might vary when the following pairs of liquids are mixed in different proportions: methanol and tetrachloromethane; water and propanol. They could also try and explain these changes.

Common learner misconceptions

- Many learners tend to think that when mixtures are formed, there is no enthalpy change. This misconception is reinforced by some junior science text books.

Answers to the workbook questions

a Please refer to completed Table 11.6

Volume of $chcl_3$ / cm³	Volume of CH_3OH / cm³	Increase in temperature (first run) / °C	Increase in temperature (second run) / °C	Average increase in temperature / °C
0	60	0	0	0
5	55	0.3	0.5	0.4
10	50	0.9	0.7	0.8
15	45	1.0	1.4	1.2
20	40	1.1	1.4	1.25
25	35	2.0	2.0	2.0
30	30	2.1	2.6	2.35
35	25	2.7	2.6	2.65
40	20	2.9	2.7	2.8
45	15	2.7	2.6	2.65
50	10	1.8	2.3	2.05
55	5	1.2	0.8	1.0
60	0	0	0	0

Table 11.6

b Dependent variable: temperature change

Independent variable: volume ratio of trichloromethane / methanol (or volume of trichloromethane)

c   Please refer to Figure 11.4

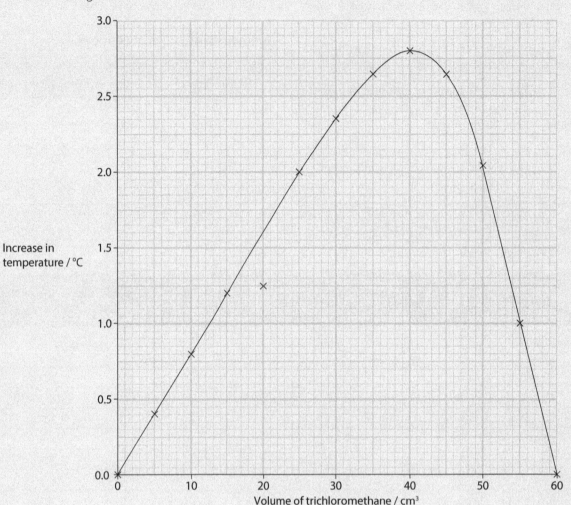

Figure 11.4

d   40 cm³ CHCl₃

e   mass (g) = density (g cm⁻³) × volume (cm³)

mass CHCl₃ = 58.8 g

mass CH₃OH = 15.8 g

f   mol = CHCl₃ = $\frac{58.8}{119.5}$ = 0.485 mol

mol = CH₃OH = $\frac{15.8}{32.0}$ = 0.49 mol

g   The highest temperature rise is when the number of moles of CHCl₃ and CH₃OH are the same. Net bond forming is exothermic. This suggests that an intermolecular 'bond' has been formed between the CHCl₃ and CH₃OH in a 1:1 ratio. This 'bonding' is stronger than the 'bonding' within either liquid alone. In CHCl₃ alone, there is dipole–dipole intermolecular bonding. In methanol alone, there is hydrogen bonding. When CHCl₃ is mixed with CH₃OH the new intermolecular forces formed are stronger than the old ones which are broken.

h   The temperature before and after mixing has not been recorded. Primary data is therefore not present. Errors in calculations may have been made. There is no evidence for correction for heat losses.

i   The point at 20 cm³ CHCl₃ is anomalous because it does not fall in line with the rest of the data. It should be ignored.

j   There is a wide range of variation in the data for some points. For example: when the volume of CHCl₃ is 30 cm³ (25% difference), 50 cm³ (28% difference) and 55 cm³ (33% difference). Although these variations have cancelled out, they are significant enough to require the experiment to be repeated several more times until consistent results are obtained.

k   Evaporation of the solvents: because they have low boiling points. Errors in measuring the volumes of liquids: because a 100 cm³ measuring cylinder is used. This does not measure accurately to the nearest 1 cm³. A burette could have been used instead. Errors in measuring the temperature accurately: because the temperature rise is very small. Very small errors are accentuated.

Chapter 12:
Electrochemistry

Chapter outline

This chapter relates to Chapter 20: Electrochemistry in the coursebook.

In this chapter, learners will complete investigations on:

- 12.1 Determining the Faraday constant
- 12.2 Comparing the voltage of electrochemical cells
- 12.3 Half-cells containing only ions as reactants
- 12.4 Planning investigation into changing the concentration of ions in an electrochemical cell
- 12.5 Planning and data analysis investigation into electrical conductivity of ethanoic aid

Practical investigation 12.1:
Determining the Faraday constant

Skills focus

The following skill areas are developed and practised (see the skill grids at the front of this guide for codes):

PI	Control experiments and identification of variables (c)
	Considering hazards (a)
COI	Methods used (c)
	Carrying out the experiment (a, b, c, d)
HI	Collecting and displaying data (b)
	Manipulating data (b)
DA	Identifying trends and patterns (a, b)
	Identifying and using calculations (a)
CP	Making predictions (a, b)
EI	Identifying problems with the procedure (a, d)
	Making a judgement on the conclusions (a)

Duration

The practical work will take at least 90 minutes. The lengthy time interval required once the experiment has been set up requires that the learners be given some relevant work to do during this period (e.g. some questions about electrolysis). The analysis and evaluation questions will take 20 minutes.

Preparing for the investigation

- Learners should have some experience of electrolysis from their previous studies.
- Before starting the experiment, learners should be aware of how to weigh a piece of metal accurately and how to use the variable resistor to control the electric current.

Equipment

Each learner or group will need:

- 0–1 A ammeter
- 100 Ohms variable resistor
- 6 V power pack or battery pack
- electrical on–off switch
- five connecting wires
- 150 cm^3 glass beaker
- cardboard electrode holder
- 100 cm^3 of 0.5 mol dm^{-3} copper(II) sulfate solution

- two copper foils 6 cm × 2 cm (for use as electrodes)
- two crocodile clips
- clock or watch to record to 40 minutes
- plastic gloves

Access to:
- distilled water in wash bottle
- 2 mol dm^{-3} nitric acid
- ethanol
- tweezers or clean tongs
- drying oven set at 100 °C
- balance to weigh to at least two decimal places

Additional notes and advice

- You may have to liaise with your physics department to obtain the variable resistors, ammeters and other electrical components. If apparatus is very scarce, the experiment can be done as a teacher demonstration with the aid of different learners doing different operations.

- 100 cm^3 of copper(II) sulfate electrolyte is prepared by dissolving 12.5 g of copper(II) sulfate-5-water in 92 g of distilled water then adding 3 cm^3 of 2 mol dm^{-3} sulfuric acid and 5 cm^3 of ethanol. This can be scaled up as appropriate for the relevant number of learners/groups doing the experiment.

- If 150 cm^3 beakers are not available, the experiment can be carried out using 250 cm^3 beakers using 150 cm^3 copper(II) sulfate.

- The electrode holder can be made from a small piece of thick cardboard about 3 cm × 5 cm with two slits to hold the copper foils in place. Alternatively, slightly longer electrodes can be used (e.g. 8 cm × 2 cm) and the top of the electrodes can be folded round the outside of the beaker.

- The clean tweezers or tongs should be placed by the nitric acid and ethanol for cleaning.

Safety considerations

- Learners should wear eye protection throughout the experiment.
- Copper(II) sulfate is harmful.
- Dilute nitric acid is an irritant.
- Ethanol is highly flammable.

- Ensure that learners hold the copper foils carefully using tongs – the edges are sharp.

Carrying out the investigation

- Learners determine the Faraday constant by measuring the gain in mass of a copper cathode during electrolysis of copper(II) sulfate solution. They then do a calculation to determine the Faraday constant.

- Because of the lengthy time interval once the experiment has been set up, some learners may forget to observe the ammeter readings at intervals and adjust the variable resistor accordingly.

- Many learners have a minimal experience of weighing materials correctly and accurately. Since a fairly sensitive balance is required, learners should be advised that air currents on the balance pan or movement of the bench on which the balance is placed should be minimised.

- Some learners may place the electrodes in the beaker at an angle so that they slide together and then touch, leading to a short circuit. This can be prevented by folding the top of each foil over the side of the beaker.

- A minimal movement of the cathode is essential since some of the copper deposited adheres poorly to the surface and some may even drop off. It is essential that at the end of electrolysis, the copper cathode is removed carefully and washed very carefully so that minimum deposit is removed. Although this experiment is better carried out by determining the loss in mass of the anode, this method is used so that learners can answer question **j**.

- Some learners may need help in setting up the apparatus and in adjusting the variable resistor. You might consider setting up the apparatus for them as well as helping them with the weighing and washing.

- Some learners may need help with part **d** of the calculation where the charge on the Cu^{2+} ion needs to be taken into account.

- More able learners could be put into groups to test the relative merits of weighing the anode instead of the cathode or be asked how the apparatus and calculation would differ if silver foils were used.

Common learner misconceptions

- A common error is to forget to convert the data into the correct units. For example, in part **a**, minutes should be converted into seconds.

Cambridge International AS & A Level Chemistry

Sample results

Table 12.1 provides results that learners could obtain in the investigation.

Initial mass of cathode	Final mass of cathode
7.63 g	7.80 g

Table 12.1

Answers to the workbook questions (using the sample results)

a $0.2 \times 45 \times 60$ C = 540 C

b $\dfrac{\text{mass Cu deposited}}{\text{Ar of copper}} = \dfrac{7.80 - 7.63}{63.5} = 2.677 \times 10^{-3}$ mol

c 2 mol

d Faraday constant = $\dfrac{\text{charge passed (a)}}{\text{mol Cu deposited (b)}} \times \dfrac{1}{2}$

   $= \dfrac{540}{5.35 \times 10^{-3}} = 100\,853$ C mol^{-1}

e Lighter pink colour on cathode; anode cleaner and lighter in colour; small amount of deposit below the cathode (generally only seen if current is too high).

f Not all the copper is deposited on anode: some may flake off or be lost in washing. So the calculated number of moles of copper is lower, thus the charge passed divided by moles of copper deposited is higher.

g Final mass minus initial mass would be too high, so the number of moles of copper deposited would be too high; so charge divided by number of moles of copper (Faraday constant) is too low.

h Nitric acid to remove impurities that would prevent the copper sticking to the cathode or being removed from the anode (nitric acid to react and remove ionic impurities; ethanol to remove residual water easily. Also allow: ethanol to remove greasy impurities.

i Reading of the ammeter was not constant. Temperature differences throughout the experiment need to be controlled (due to heating effect of current). Removal of copper from cathode during washing leads to errors.

j Not all copper was deposited on cathode (some may flake off or be lost in washing).

Practical investigation 12.2: Comparing the voltage of electrochemical cells

Skills focus

The following skill areas are developed and practised (see the skill grids at the front of this guide for codes):

COI Methods used (c)
     Carrying out the experiment (a, b, c, d)
HI   Manipulating data (b)
DA   Identifying trends and patterns (a)
CP   Drawing conclusions (e)
EI   Identifying problems with the procedure (a, e)
     Making a judgement on the conclusions (a)

Duration

The practical work will take 30 minutes. The analysis and evaluation questions will take 15 minutes.

Preparing for the investigation

- Before starting the experiment, learners should be aware of the terms used in describing electrochemical cells, for example, half-cell, salt bridge, polarity.

- Before doing the investigation, it is useful to demonstrate the reaction of zinc with copper(II) sulfate and zinc with iron(II) sulfate, and the reaction of powdered iron with copper(II) sulfate. This serves to demonstrate the relative reactivity of zinc, copper and iron.

Equipment

Each learner or group will need:

- high resistance voltmeter
- 100 cm^3 three beakers
- two connecting wires
- two crocodile clips
- three strips of filter paper 10 cm × 1 cm soaked with saturated potassium nitrate solution

Chapter 12: Electrochemistry

- emery paper (or sandpaper)
- iron nail
- zinc foil, 6 cm × 2 cm
- copper foil, 6 cm × 2 cm
- 50 cm³ 1.0 mol dm⁻³ aqueous copper(II) sulfate
- 50 cm³ 1.0 mol dm⁻³ aqueous zinc sulfate
- 50 cm³ 1.0 mol dm⁻³ aqueous acidified iron(II) sulfate
- plastic gloves
- distilled water in wash bottle and paper towels

Additional notes and advice:

- You may have to liaise with your physics department to obtain the voltmeters and other electrical components.
- A large, old-fashioned iron nail works best. Shiny modern nails do not work as well.

Safety considerations

- Learners should wear eye protection throughout the experiment.
- Copper(II) sulfate and 1.0 mol dm⁻³ solutions of copper(II) sulfate are harmful and also irritants. The same applies to both solid and 1.0 mol dm⁻³ solutions of zinc sulfate and iron(II) sulfate.

Carrying out the investigation

- Learners compare the voltages of three electrochemical cells and predict the polarity of each of the half-cells involved.
- Learners should ensure that the copper, zinc and iron nails are as clean as possible. The presence of grease or dirt on the electrodes may lead to incorrect voltages being recorded.
- Encourage learners not to let the crocodile clips dip into the solution since these may alter the voltage slightly.
- The zinc sulfate and iron(II) sulfate solutions are both colourless. Ensure that learners label these solutions.
- Some learners may need help in recording a steady reading on the voltmeter.
- Others may need help in determining the negative and the positive pole of the cell.

More able learners could use books or the internet to find out about the use of 'cell diagrams' as a shorthand way of representing electrochemical cells.

Common learner misconceptions

- Some learners may confuse the idea of loss and gain of electrons when referring to reducing agents and oxidising agents. The more reactive the metal, the better it releases electrons and the better it acts as a reducing agent.

Sample results

Table 12.2 provides results the learners may obtain in the investigation.

		E_{cell} values
Theoretical	zinc / copper cell	+ 1.10 V
	zinc / iron cell	+ 0.32 V
	iron / copper cell	+ 0.78 V
Experimental	zinc / copper cell	+ 1.0 V
	zinc / iron cell	+ 0.20 V
	iron / copper cell	+ 0.6 V

Table 12.2

Answers to the workbook questions (using the sample results)

a 0.8 V

b Cell A zinc negative and copper positive; cell B zinc negative and iron positive; cell C iron negative and copper positive

c Zinc: it is highest in the electrochemical series / it releases electrons better than the other metals.

d To remove oxide layer or rust on the surface.

e So that there is no contamination from the previous solutions.

f Experimental is value lower than predicted. Iron was not pure / steel was used instead of iron. The Fe(II) sulfate solution may oxidise in the air. Also allow: iron nail has different a surface area to the copper foil.

g Temperature may change during the experiment. Temperature is not a constant standard 25 °C.

Practical investigation 12.3:
Half-cells containing only ions as reactants

Skills focus
The following skill areas are developed and practised (see the skill grids at the front of this guide for codes):

PI	Selecting information (c)
	Defining the problem under investigation (d)
COI	Methods used (a, c, d)
	Carrying out the experiment (a, b, c, d)
HI	Collecting and displaying data (b)
CP	Drawing conclusions (b)
EI	Making a judgement on the conclusions (a)

Duration
The practical work will take 25 minutes. The analysis and evaluation questions will take 30 minutes.

Preparing for the investigation
- Before starting the experiment, learners should be able to construct half equations, understand redox reactions in terms of electron transfer and understand the meaning of the word 'standard' as applied to electrochemical cells.
- Before doing the investigation it is useful to demonstrate the use of a platinum electrode in a half-cell involving only metal ions or metal ions and molecules.

Equipment
Each learner or group will need access to:
- six test tubes in a test-tube rack
- dropping pipettes
- a beaker for distilled water to wash pipettes if needed

Access to:
- distilled water
- 0.1 mol dm^{-3} aqueous ammonium iron(III) sulfate (iron alum)
- 0.1 mol dm^{-3} aqueous potassium iodide
- 1% starch solution in a small bottle with a dropping pipette
- 1% aqueous potassium hexacyanoferrate(III) solution in a small bottle with a dropping pipette

Safety considerations
- At the concentrations used, all the solutions are low risk although potassium hexacyanoferrate(III) may cause eye irritation and skin irritation.

Carrying out the investigation
- Learners investigate the reaction between aqueous iron(II) ions and iodide ions to determine the products formed using aqueous potassium hexacyanoferrate(III) to test for iron(II) ions and starch solution to test for iodine. Learners are then asked to plan the construction of an electrochemical cell for this reaction.
- Care should be taken to avoid cross contamination of the solutions. Ensure that learners only use the dropping pipettes provided with each test solution.
- Some learners may need help in constructing the ionic equation in part **b**.
- Other learners may need help with planning the construction of the electrochemical cell.
- More able learners could devise more complex half-cells, for example, for reactions involving acidified potassium manganate(VII).
- Some could also help other learners with planning the construction of the electrochemical cell once their own answers have been checked.

Common learner misconceptions
- Some learners incorrectly suggest that a reactive metal, instead of platinum, should be used in a half-cell containing only metal ions. For example, using an iron electrode in a solution of Fe^{2+} and Fe^{3+} ions.

Sample results
Table 12.3 provides results the learners may obtain in the investigation.

Method	
Step 1	colourless solution / no change
Step 2	yellow solution / no change
Step 3	colourless solution / no change
Step 4	colourless solution / no change
Step 6	solution turns brown
Step 8	solution turns blue black
Step 9	solution turns dark blue

Table 12.3

Chapter 12: Electrochemistry

> **Answers to the workbook questions (using the sample results)**
>
> a Iron(II) ions, Fe^{2+}, iodine, I_2
>
> b $Fe^{3+}(aq) + e^- \rightarrow Fe^{2+}(aq)$ and $2I^-(aq) \rightarrow I_2(aq) + 2e^-$
>
> c $2Fe^{3+}(aq) + 2I^-(aq) \rightarrow 2Fe^{2+}(aq) + I_2(aq)$
>
> d I^- is a better reducing agent than Fe^{2+} and Fe^{3+} is a better oxidising agent than I_2
>
> e Similar drawing to Figure 12.1 in the workbook including salt bridge and voltmeter:
>
> the solution in one beaker is 1 mol dm^{-3} with respect to $Fe^{3+}(aq)$ and 1 mol dm^{-3} with respect to $Fe^{2+}(aq)$; the solution in the other beaker is 1 mol dm^{-3} with respect to $I^-(aq)$ and 1 mol dm^{-3} with respect to $I_2(aq)$.
>
> Pt electrodes should be dipping into each solution.
>
> f To ensure that the reactants did not react with the test reagents.
>
> g The temperature should be 25 °C; the concentration of ions should be 1.0 mol dm^{-3}.

Practical investigation 12.4: Planning

Investigation into changing the concentration of ions in an electrochemical cell

Skills focus

The following skill areas are developed and practised (see the skill grids at the front of this guide for codes):

PI	Defining the problem under investigation (c, d) Control experiments and identification of variables (a, b)
COI	Methods used (c) Carrying out the experiment (a, b, c, d)
HI	Collecting and displaying data (b) Manipulating data (b, c)
DA	Identifying trends and patterns (a, c) Identifying and using calculations (b)
CP	Drawing conclusions (b)
EI	Making a judgement on the conclusions (a)

Duration

The planning will take 30 minutes (or 90 minutes if practical work is included); the analysis and evaluation questions will take 30 minutes. The experiment does not always give good results and you may consider this as a planning exercise only. The preparation of the solutions of different concentrations is the most time-consuming step. You may consider preparing the solutions of copper(II) sulfate of different concentrations beforehand.

Preparing for the investigation

- Learners should already have experience of setting up the equipment in investigation 12.2.
- It is very important to pay attention to the cleanliness of the copper half-cells as small amounts of impurities will give rise to large errors.

Equipment

Each learner or group will need:

- high resistance voltmeter
- four 100 cm³ beakers
- two connecting wires
- two crocodile clips
- two 100 cm³ glass beakers
- four strips of filter paper, 10 cm × 1 cm, soaked with saturated potassium nitrate solution

Access to:

- distilled water and paper towels
- burette (for distilled water)
- funnel for filling burette
- two strips of copper foil, 6 cm × 2 cm
- a strip of zinc foil, 6 cm × 2 cm
- 100 cm³ of 1.0 mol dm^{-3} copper(II) sulfate solution
- 50 cm³ of 1.0 mol dm^{-3} zinc sulfate solution

- plastic gloves
- emery paper (or sandpaper)
- burette or graduated pipette with funnel

Method

A suggested method is:

1. Clean the strips of copper and zinc with emery paper or sandpaper.
2. Set up the apparatus (as in Practical investigation 12.2) with a zinc half-cell and a copper half-cell containing 1.0 mol dm^{-3} copper(II) sulfate solution.
3. Connect the two half-cells with a salt bridge made from a strip of filter paper soaked in potassium nitrate solution and connect to the external circuit.
4. Record the steady voltage on the voltmeter.
5. Remove the strip of copper and wash it with distilled water then dry it with a paper towel.
6. Use a burette or graduated pipette to make up a solution of copper(II) sulfate of concentration 0.5 mol dm^{-3} by taking 25 cm^3 of 1.0 mol dm^{-3} copper(II) sulfate solution and adding 25 cm^3 distilled water from a burette.
7. Repeat Steps 3 and 4 using copper(II) sulfate solution of concentration 0.5 mol dm^{-3}. Use a fresh salt bridge.
8. Repeat the experiment using copper(II) sulfate solutions of concentration 0.1 mol dm^{-3} then 0.01 mol dm^{-3} and finally 0.001 mol dm^{-3}.

Safety considerations

- Learners should wear eye protection throughout the experiment.
- Copper(II) sulfate and 1.0 mol dm^{-3} solutions of copper(II) sulfate and zinc sulfate are harmful and also irritants.
- The edges of the metal foils are sharp.

Carrying out the investigation

- Learners may not have experience of serial dilution techniques; for example, to produce a 0.01 mol dm^{-3} solution of X from a 0.1 mol dm^{-3} solution of X, you take 1 cm^3 of the 0.1 mol dm^{-3} solution of X and make this up to 10 cm^3 with distilled water and then repeat this process with the 0.01 mol dm^{-3} solution to make a 0.001 mol dm^{-3} solution.
- This investigation is an extension of investigation 12.2. In this investigation, the left-hand half-cell remains a standard zinc / zinc sulfate half-cell while the right-hand half-cell is copper(II) ions of varying concentrations. The purpose of the experiment is to see how decreasing the concentration of copper(II) ions affects the value of the E_{cell}.
- Encourage learners to label their copper(II) sulfate solutions with the appropriate concentration.
- The differences in voltages are very small so it is important that learners pay attention to the accurate reading of the voltages. The results may show a scatter of data which does not show the correct relationship.
- Some learners may need help with the serial dilutions and you might consider checking that they are using the correct volumes for the dilutions. You might consider giving some learners pre-prepared solutions but telling them how they were made.
- Other learners may need help in plotting the graph in part **b**.
- More able learners could use books or the internet to find out about the use of 'cell diagrams' as a shorthand way of representing electrochemical cells.

Common learner misconceptions

- Some learners may need help in recording a steady reading on the voltmeter.

Sample results

Table 12.4 provides sample results learners may obtain in the investigation.

[Cu^{2+}] / mol dm^{-3}	log$_{10}$[Cu^{2+}]	E_{cell} / V
1.0 mol dm^{-3}		E_{cell} / V = 1.10 V
0.5 mol dm^{-3}		E_{cell} / V = 1.09 V
0.1 mol dm^{-3}		E_{cell} / V = 0.81 V
0.01 mol dm^{-3}		E_{cell} / V = 0.51 V
0.001 mol dm^{-3}		E_{cell} / V = 0.22 V

Table 12.4

Chapter 12: Electrochemistry

> **Answers to the workbook questions (using the sample results)**
>
> **a** Dependent variable is the voltage, E_{cell}. Independent variable is the concentration of Cu^{2+} ions.
>
> **b** The value of E_{cell} shows positive linear correlation with the $\log_{10}[Cu^{2+}]$
>
> **c** The value of $\log_{10}[0.05]$ is −1.30 and this should give a value of E_{cell} of 0.72 V
>
> **d** A burette has an accuracy of reading to ± 0.05 cm³, whereas a measuring cylinder has an accuracy of far less. So errors in measurements are less with a burette.
>
> **e** Serial dilution can lead to errors which are unpredictable. They may cancel each other out because the error of the first dilution may lead to a concentration which is too low and the error in the second may lead to a concentration which is too high. If a consistent error is made, for example, the dilution is too great each time, the effect will be magnified with each dilution.
>
> **f** Weigh out a small calculated amount of copper(II) sulfate into a weighing bottle using a balance to two or three decimal places. Transfer the copper(II) sulfate from the weighing bottle to a small beaker and reweigh the weighing bottle. Dissolve the copper(II) sulfate in a small amount of distilled water in a beaker. Transfer the solution to a 1.0 dm³ volumetric flask and wash out the beaker several times with distilled water and transfer to the flask. Make up the volume in the flask to the graduation line and shake.

Practical investigation 12.5: Planning and data analysis

Investigation into electrical conductivity of ethanoic aid

Extension investigation

Skills focus

The following skill areas are developed and practised (see the skill grids at the front of this guide for codes):

PI	Selecting information (c)
	Defining the problem under investigation (a, c, d)
	Control experiments and identification of variables (c)
	Considering hazards (a, b)
COI	Methods used (a, b, d)
	Carrying out the experiment (a, b, c, d)
HI	Collecting and displaying data (b)
	Manipulating data (b, c)
DA	Identifying trends and patterns (a)
CP	Drawing conclusions (b)
EI	Identifying problems with the data (a)
	Making a judgement on the conclusions (a)

Duration

The planning and data analysis and evaluation questions will take 45 minutes.

Method

A suggested method is:

1. Make up dilute solutions of ethanoic acid and sodium ethanoate of stated concentration. Distilled water or deionised water should be used. The concentrations should be equimolar and dilute enough to minimise any hazards, for example, 1.0 mol dm⁻³. Learners should stress that the glassware used should be very clean. They should make a comment about the hazards associated with making up a solution of ethanoic acid, for example, carry out away from naked flames and wear gloves (and eye protection) when making up the solution.

2. Allow the solutions and the water used to make these up to come to the same temperature. Learners could comment about conducting the experiment in a water bath (at 25 °C).

3. Clamp the conductivity cell so that it dips into the water used to make up the solution. Keep the water stirred and read the meter until a constant value is obtained. Record this value.

4. Wash and dry the conductivity cell then repeat Step 3 using a (1.0 mol dm⁻³) solution of ethanoic acid instead of water under the same conditions.

5. Wash and dry the conductivity cell then repeat Step 3 using a (1.0 mol dm⁻³) solution of sodium ethanoate under the same conditions.

6. Repeat results until they are consistent.

Carrying out the investigation

- The information needed for this investigation is given in the workbook.

- The learners are asked to plan an experiment to compare the electrical conductivity of ethanoic acid and sodium ethanoate and analyse data about the electrical conductivity of ethanoic acid.

  Some learners may need help with planning the investigation in sufficient detail. You may need to give learners hints about the control experiment and other details such as keeping the solutions stirred and the importance of temperature control, for example, by referring to the rate of molecular movement with temperature.

- The symbol for the units of resistance, Ω, may have to be explained to those candidates who do not do physics, as well as the idea that the resistance is inversely related to the conductance.

- Some candidates may need help in determining the units for molar conductivity. Reference to how units are determined from rates of reaction or equilibrium constants should help them.

  Learners who finish early can compare the relative conductivity of various ionic compounds and determine whether there are any patterns relating to the size of the ions or the charge of the ions.

Common learner misconceptions

- Many learners think that distilled water does not contain ions. Ions arise from the glassware in the distillation process and so it is necessary to do a control experiment with the water alone to account for these.

Answers to the workbook questions (using the sample results)

a $\Omega^{-1}\,mol^{-1}\,m^2$. Also accept $S\,mol^{-1}\,m^2$, as S (siemens) is the SI unit.

b Please refer to Figure 12.1.

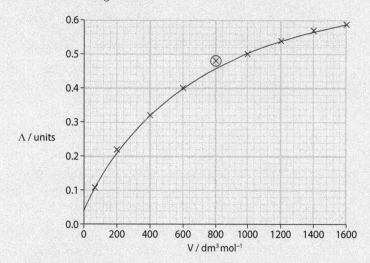

Figure 12.1

c Anomalous point (0.48) circled. The point is anomalous as it does not fall on the steady curve; it does not fit the general pattern. The point was ignored when the line was drawn.

d Ethanoic acid becomes more ionised as the dilution increases. The extent of increase of ionisation is relatively smaller with increasing dilution.

e Conductivity is proportional to area and inversely proportional to distance.

f Temperature: increase in temperature make ions move faster.

  Concentration: increase in concentration increases the number of ions (which conduct).

g Ions are present in both tap water and distilled water in varying amounts.

h Traces of (ionic) impurities may conduct electricity.

i Plastic (polypropene) because it does not (usually) contain ions.

Chapter 13:
Further aspects of equilibria

Chapter outline

This chapter relates to Chapter 21: Further aspects of equilibria and Chapter 23: Entropy and Gibbs free energy in the coursebook.

In this chapter learners will complete investigations on:

- 13.1 Change in pH during an acid–base titration
- 13.2 Data analysis investigation into partition of ammonia between water and trichloromethane
- 13.3 Planning investigation into an esterification reaction at equilibrium
- 13.4 Planning investigation into the effect of temperature on the $N_2O_4 \rightleftharpoons 2NO_2$ equilibrium
- 13.5 Data analysis investigation into equilibrium entropy and enthalpy change

Practical investigation 13.1:
Change in pH during an acid–base titration

Skills focus

The following skill areas are developed and practised (see the skills grids at the front of this guide for codes):

HI	Collecting and displaying data (b, c)
	Manipulating data (a, b, c)
DA	Identifying trends and patterns (a)
	Identifying and using calculations (a)
CP	Drawing conclusions (b)
EI	Identifying problems with the procedure (a)
	Identifying problems with the data (a)

Duration

- The practical work will take about 40 minutes but may be shortened by suggesting an initial addition of 10 cm³ of the aqueous sodium hydroxide.
- The analysis and evaluation questions will take 30 minutes.

Preparing for the investigation

- Learners should have some experience of titrations from their previous studies and know how to manipulate a burette and a volumetric pipette.
- They should also know how to carry out calculations involving moles, volumes and concentrations.
- Learners should be instructed in the careful use of the glass pH electrode if one is used.
- It is useful if the table of results is drawn up prior to carrying out experimental work.

Equipment

Each learner or group will need:

- 25 cm³ volumetric pipette
- pipette filler
- 50 cm³ burette
- 100 cm³ beaker
- glass stirring rod or magnetic stirrer
- pH meter and pH electrode

- two clamps and clamp stand for burette and pH electrode
- funnel to fill burette

Access to:
- dilute ethanoic acid of unknown concentration
- 0.10 mol dm^{-3} sodium hydroxide

Additional notes and advice

If insufficient pH meters are available, the teacher could carry out the investigation with learners taking turns to add the aqueous sodium hydroxide from the burette or reading the pH meter.

Safety considerations

- Learners should wear eye protection throughout the experiment.
- Warn learners that sodium hydroxide is an irritant at a concentration of 0.10 mol dm^{-3}.
- When making up the solutions, technicians should be aware that glacial ethanoic acid is corrosive and flammable, concentrated ethanoic acid is corrosive and solid sodium hydroxide is corrosive.

Carrying out the investigation

- If a magnetic stirrer is used, ensure that it is not rotating too rapidly and that it will not hit the side of the glass pH electrode. Learners should also be advised not to stir too forcefully with the glass rod.
- The tip of the burette, the pH electrode and the glass stirring rod all need to be placed within a small area in the beaker. This tight fit means that it is essential that the burette tip is at one side of the beaker and the pH electrode is at the opposite side.
- Ensure that sufficient time is left after each addition of sodium hydroxide for the pH electrode to respond to the change in pH after stirring. Ten seconds should be sufficient. In some old pH meters the pH may not appear to stay constant. In these cases, an average pH should be assessed.
- Some learners will need help in setting up the apparatus, given that the tip of the burette, the pH electrode and the glass stirring rod all need to be within the beaker. It is better that a single clamp stand with two clamps attached is used rather than two separate clamp stands.
- Some learners have minimal experience of using a volumetric pipette and burette. You may need to help them deliver the correct volume of acid from the pipette.
- More able learners could calculate the pH at the end-point of a strong acid–strong base titration by reference to a standard drop size of 0.05 cm^3. For example, for 0.1 mol dm^{-3} hydrochloric acid added to 25 cm^3 of 0.1 mol dm^{-3} sodium hydroxide:

Concentration of H$^+$ ions one drop before the end-point =
$\frac{0.05 \times 0.1}{49.95}$ = 1.0 × 10^{-4} mol dm^{-3} so pH = 4

Concentration of OH$^-$ ions one drop after the end-point =
$\frac{0.05 \times 0.1}{51.05}$ = 1.0 × 10^{-4} mol dm^{-3}

K_w = [H$^+$][OH$^-$] so [H$^+$] = $\frac{10^{-14}}{1.0 \times 10^{-14}}$ = 1.0 × 10^{-10} so pH = 10

Sample results

These give an idea of some results the learners may obtain in the investigation.

Volume of NaOH /cm^3	pH
0	3.6
2	4.0
4	4.6
6	4.5
8	4.5
10	4.7
12	4.7
14	5.0
16	5.0
18	5.2
20	5.4
22	5.9
24	6.7
26	10.8
28	11.2
30	11.4
32	11.6
34	11.7

Table 13.1

Answers to the workbook questions (using the sample results)

a See Figure 13.1

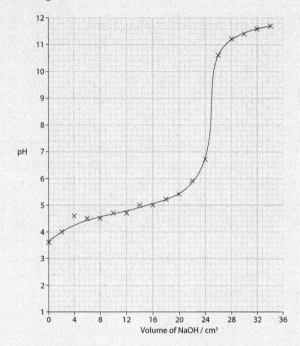

Figure 13.1

b The curve is S-shaped. There is a relatively small increase in pH until about 22 cm³ of sodium hydroxide is added. There is rapid change between 22 cm³ and 26 cm³ of added sodium hydroxide. After 26 cm³ of sodium hydroxide has been added, there is a relatively small increase in pH again. The end-point of the titration coincides with the most rapid change in pH.

c 24.8 cm³. It is half way along the steepest part of the curve.

d A neutral solution is pH 7. The end-point is approximately pH 8.5 which is alkaline.

e Moles sodium hydroxide = $\frac{24.8}{1000} \times 0.1 = 2.48 \times 10^{-3}$ mol

Moles ethanoic acid = 2.48×10^{-3} mol (since one mol of ethanoic acid reacts with one mol of sodium hydroxide)

Concentration of ethanoic acid = $2.48 \times 10^{-3} \times \frac{1000}{25} = 0.10$ mol dm⁻³ (to two significant figures).

f Circle around the value at 4 cm³ sodium hydroxide added. The point does not fall on the smooth curve. The pH may have been recorded before the mixture was stirred, so the pH was too high. The anomalous point should be ignored.

g When the pH change starts to increase more rapidly (e.g. the increase is more than 0.5 of a pH unit), sodium hydroxide should be added in smaller volumes (e.g. 0.05 cm³).

h Removing them would also remove some of the solution. Therefore, the number of moles of acid titrated would be reduced. If they were only removed once, the amount removed would be very small compared with the overall volume of the solution, so it would not make much difference to the end result. If they were removed after each addition of sodium hydroxide, the amount removed might make a difference to the end result.

i Use a larger beaker and greater volumes to allow easier stirring and addition of sodium hydroxide. Leave a longer time between adding a portion of sodium hydroxide and recording the pH to allow better equilibration. Repeat the experiment until consistent results are obtained.

Practical investigation 13.2: Data analysis
Investigation into partition of ammonia between water and trichloromethane

Skills focus

The following skill areas are developed and practised (see the skills grids at the front of this guide for codes):

PI	Defining the problem under investigation (c)
	Control experiments and identification of variables (c)
	Considering hazards (b)
COI	Methods used (c)
	Carrying out the experiment (a, b, c, d)
HI	Manipulating data (b, c)
DA	Identifying trends and patterns (a)
CP	Drawing conclusions (b)
EI	Identifying problems with the data (c)
	Making a judgement on the conclusions (a)

Duration

Graph plotting and answering questions will take 30–40 minutes.

Preparing for the investigation

- Learners should have a basic understanding of the concept of the partition of a solute between two immiscible solvents.

- They should also be able to understand the equilibrium aspects of partition coefficients.

Equipment

- It is not advised that an experiment be conducted because trichloromethane is harmful and ammonia vapour is toxic.

Safety considerations

- Learners should answer the safety consideration questions given the information in the workbook. For example, ammonia is a gas at room temperature and may evaporate from the solutions. Trichloromethane has a low boiling point so is volatile. The experiment should therefore be carried out in a fume cupboard.

Carrying out the investigation

- The learners are asked to analyse data about the equilibrium concentrations of ammonia in water and trichloromethane when the solutions are diluted.

- Some learners may need help in using the full extent of the graph paper and with drawing the line which curves only at very low concentrations.

- Other learners may need help in answering part **a**, where the equilibrium constant must be the same for each dilution.

- Some learners may need help in answering part **f**, where they may think that any points not on the predominantly straight line will be anomalous.

- Learners could be asked to suggest a reason for the deviation in the line at low concentration. They could be given a hint to consider the equation for the equilibrium reaction of ammonia with water and the consideration that both ammonia and trichloromethane are simple molecules which do not react. The deviation from the experimental curve can be explained by the partial ionisation of ammonia in water. In a 1.0 mol dm^{-3} solution, approximately one ammonia molecule in 10 000 is present as NH_4^+ ions and OH^-. This proportion increases with dilution.

Common learners misconceptions

- Some learners will think that the addition of an equal volume of both water and trichloromethane to the mixture will shift the equilibrium to the right or left; they may think that the graph they draw represents this shift.

Answers to the workbook questions (using the sample results)

a Straight line of increasing gradient going through the origin – proportional relationship. No matter how much the concentration in the water layer and the trichloromethane layer have been diluted, the ammonia will always move from one layer to the other until the equilibrium is achieved.

b See Figure 13.2

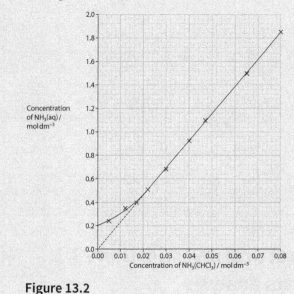

Figure 13.2

Chapter 13: Further aspects of equilibria

c   Any points on the graph can be used (e.g. 1.62 on the vertical axis and 0.07 on the horizontal axis). $\frac{1.62}{0.07}$ = 23.1. Alternatively, the gradient of the graph will give the same result.

d $\frac{0.32 - 0.24}{0.24} \times 100 = 33\%$

e   A glass stopper for the separating funnel to allow mixing of the layers by shaking without loss of liquid. This would also reduce the amount of evaporation of ammonia from the water because ammonia is a gas at room temperature (it has a low boiling point).

f   There is a consistent trend in the values at low concentrations.

g   Do a greater range of dilutions at low concentrations and/or repeat the experiments at the lower concentrations.

h
- Sufficient experiments of the same type (just at different concentrations) give the same equilibrium constant
- There may be other errors which have not been accounted for.

i   Temperature – ammonia has a different solubility at different temperatures. More ammonia may come out of solution during the titration at a high temperature than a low temperature. Also allow – less ammonia dissolves at lower pressures.

Practical investigation 13.3: Planning

Investigation into an esterification reaction at equilibrium

Skills focus

The following skill areas are developed and practised (see the skills grids at the front of this guide for codes):

PI   Defining the problem under investigation (a, c, d)
     Control experiments and identification of variables (c)
     Considering hazards (a, b)

COI Methods used (a, b, d)
     Carrying out the experiment (b)

HI   Collecting and displaying data (b)
     Manipulating data (b, c)

DA   Identifying trends and patterns (a)
     Identifying and using calculations (b)

CP   Drawing conclusions (e)

EI   Identifying problems with the procedure (a)
     Identifying problems with the data (b)

Duration

The planning and data analysis and evaluation questions will take at 50–70 minutes depending on how much help is needed.

Preparing for the investigation

- Learners should have a basic understanding of the principles of equilibrium and know how to deduce an equilibrium constant from relevant data.

Carrying out the investigation

- The information needed for this investigation is given in the workbook.

- Learners are asked to plan a series of experiments to determine the concentrations of ethanoic acid, ethanol and ethyl ethanoate at equilibrium. They are then asked to analyse data about the concentrations of each of these substances at equilibrium in order to determine the equilibrium constant.

- Some learners may need help with planning the investigation in sufficient detail. You may need to give learners hints about keeping the reaction mixture in a closed container to prevent evaporation of the volatile components. Hints about other details, such as keeping the solutions stirred and the importance of temperature control may also need to be given.

- Some learners may need help in selecting suitable volumes or masses of ethyl ethanoate, water and catalyst. Their attention may need to be drawn to relationship density = mass/volume.

- Although a hint about the analysis of the acid in the equilibrium mixture by titration is given in the workbook, some learners may need help in selecting a suitable indicator for the titration of a weak acid with a strong base. Others may forget to account for the hydrochloric acid catalyst.

- Some may need help in writing the equilibrium constant and calculating the amounts of ethanoic acid and ethanol at equilibrium by subtraction.

🛈 Learners who finish early could be asked to devise a practical procedure for another equilibrium reaction, for example, $H_2 + I_2 \rightleftharpoons 2HI$ using sodium thiosulfate to determine the amount of iodine.

Common learners misconceptions

- Many learners confuse the ideas of equilibrium and rate of reaction, or they believe that the equilibrium constant changes when the concentrations of individual substances in the equilibrium mixture changes.

Method

Suggested answers to the questions about method include:

1 The amounts chosen should be such that the acid could give a titre which is not too small or not too large. Suitable proportions could be 20 cm³ of ethyl ethanoate + 2.0 cm³ of hydrochloric acid + 3.0 cm³ water. At least four different proportions of ethyl ethanoate and water should be suggested but the volume of hydrochloric acid should remain the same. The total volume should remain the same (e.g. 15 cm³ of ethyl ethanoate + 2.0 cm³ of hydrochloric acid + 8.0 cm³ water or 10 cm³ of ethyl ethanoate + 2.0 cm³ of hydrochloric acid + 13.0 cm³ water).

2 Care should be taken with the hydrochloric acid. It should be dispensed by burette or a pipette with a pipette filler. Although 6.0 mol dm⁻³ hydrochloric acid is an irritant according to some safety sources, other sources regard it as a corrosive substance. Safety clothing such as gloves and eye protection are required. The ethyl ethanoate should be kept away from naked flames since it is flammable. The products of the reaction are also flammable but, since they are in solution, there is less danger. It is better if the titration is carried out in a well-ventilated room.

3 Suggested steps:

a Four different flasks are set up with different proportions of ethyl ethanoate, hydrochloric acid and water.

b The liquids are dispensed from separate burettes or graduated pipettes.

c The flasks are stoppered with a glass stopper to prevent ethyl ethanoate vapour reacting with rubber or plastic. The flask should be stoppered to prevent loss of ethyl ethanoate, ethanol or ethanoic acid vapour.

d The flasks are left for one week at constant temperature with occasional shaking to mix the liquids.

e After one week, a sample of known volume is taken from each flask and placed in four separate titration flasks.

f A suitable indicator is added to each flask (e.g. phenolphthalein).

g The contents of each flask are titrated with aqueous sodium hydroxide of known concentration and the total volume of acids calculated.

h A separate sample of the 6.0 mol dm⁻³ hydrochloric acid (2.0 cm³ + 23.0 cm³ water) is titrated to determine the amount of acid catalyst.

i The amount of sodium hydroxide used in this titration is subtracted from the other results to allow for the acid added as a catalyst.

Chapter 13: Further aspects of equilibria

Answers to the workbook questions (using the sample results)

a See Table 13.2

Moles of CH_3COOH at start	Moles of C_2H_5OH at start	Moles of $CH_3COOC_2H_5$ at equilibrium	Moles of CH_3COOH at equilibrium	Moles of C_2H_5OH at equilibrium	K_a
1.00	0.18	0.17	0.83	0.01	3.48
1.00	0.50	0.42	0.58	0.08	3.80
1.00	0.33	0.29	0.71	0.04	2.96
1.00	2.00	0.84	0.16	1.16	3.80
1.00	8.00	0.96	0.04	7.04	3.27

Table 13.2

b In the equation for the reaction an equal number of moles of ethyl ethanoate and water are formed and neither of these were present at the start of the reaction.

c $K_c = \dfrac{[CH_3COOC_2H_5][H_2O]}{[CH_3COOH][C_2H_5OH]}$

d There are the same number of concentration terms at the top and bottom of the equilibrium expression, so the volumes cancel.

e See Table 13.2

f To prevent the loss of vapour or to prevent loss of reactants and products

g Temperature. Use an oil bath which has a thermostat, use electrical heating with controlled current or use an electrical oven with thermostat. (Although K_p and K_c vary with temperature, the effect of temperature on this reaction is small since the enthalpy change is close to zero.)

h Cool the tubes rapidly to a low temperature then break them in a titration flask (and titrate immediately) or break the tubes in a known volume of cold water (and titrate immediately).

i The data are all of the same order of magnitude and the spread of data is within the boundaries of the experimental error expected. Significant experimental errors are expected because of loss of vapours during the titration. This occurs to different extents because of the different compositions of the reaction mixtures. Some of the reaction mixture may remain in the glass tube after breaking (unless broken in water). The temperatures have not been controlled sufficiently.

Also allow – comments referring to the fact that 2.96 is significantly different to 3.80 so more experiments are needed until there are more consistent results.

j The reaction **reaches equilibrium** faster at higher temperatures and there is no need for an added catalyst.

Practical investigation 13.4: Planning

Investigation into the effect of temperature on the $N_2O_4 \rightleftharpoons 2NO_2$ equilibrium

Skills focus

The following skill areas are developed and practised (see the skills grids at the front of this guide for codes):

PI	Selecting information (c)
	Defining the problem under investigation (a, c, d, e)
	Considering hazards (a, b)
COl	Methods used (a, b, d)
HI	Collecting and displaying data (b)
CP	Making predictions (a, b)
EI	Making a judgement on the conclusions (a)

Duration

The planning will take 30 minutes and the analysis and evaluation questions will take 20 minutes.

Preparing for the investigation

- Learners should have some experience of setting up apparatus for measuring gas volumes and doing calculations involving the molar gas volume. It would be helpful beforehand to give a lesson on drawing chemical apparatus.
- Learners could suggest an equipment list and do some of the planning for prep. They could also research how acidic gases can be absorbed.

Equipment

No equipment is needed since the learners should not undertake the experiments themselves. If the teacher wishes to demonstrate the experiment after the learners have done the investigation, the apparatus for the production of nitrogen dioxide is shown in the diagrams in the method section.

The experiment must be done in a well-ventilated fume cupboard.

Additional equipment required for the teacher to demonstrate the effect of temperature on the equilibrium is given here.

- large beaker, 2 dm³, or other large transparent container capable of being heated
- gas syringe containing about 80 cm³ of nitrogen dioxide fitted at the end with a glass tap for closure

Access to:

- water cooled to about 5 °C with ice
- water warmed in a kettle to about 50°
- beaker containing 100 cm³ of concentrated sodium hydroxide solution to dispose of nitrogen dioxide from the syringe

Safety considerations

- This investigation is a planning exercise and should not be done practically since nitrogen dioxide is produced in the reaction and may not always be contained in the apparatus; it could be present in sufficient quantity to be a hazard to health.
- Learners should answer the questions on safety given the information in the workbook. The experiment for the production of dinitrogen tetroxide should allow for some air to escape from the apparatus as it is heated (as shown in the diagram). A completely closed system may lead to the stoppers being blown out and the dispersal of lead nitrate. The experiment for the formation of nitrogen dioxide should be a closed system with no leaks so that no nitrogen dioxide escapes. The experiment should be carried out in a fume cupboard to minimise any poisonous gas getting into the laboratory. The lead nitrate is toxic so should be weighed out using gloves and making sure that no dust is raised. Learners may suggest using concentrated aqueous sodium hydroxide to absorb any nitrogen dioxide. This is corrosive so care is needed dispensing it. Learners could also look up information from CLEAPSS A level cards for prep.

Carrying out the investigation

- Learners are asked to plan an experiment to prepare and collect a sample of nitrogen dioxide safely and then use this sample to demonstrate the effect of temperature on the $N_2O_4 \rightleftharpoons 2NO_2$ equilibrium.

  Some learners may need help in drawing diagrams of the apparatus. The use of a ruler or a suitable stencil may help here.

- Most learners will need help in understanding the importance of allowing or stemming the flow of gas by the use of glass taps. This could be introduced in terms of the use of a burette tap to stop or start the flow of liquid. Learners will not be expected to explain the complete sequence of removing air from the apparatus but they should have some idea of doing this.

- Others may have difficulty in organising the sequence of steps to be followed in the method section of the experiment. They could be encouraged to plan the sequence on a separate sheet of paper first and then rearrange as necessary.

- Some learners may need help in answering parts **b** and **c**, where mole calculations are required.

  Learners could be asked to suggest an experiment to demonstrate how pressure affects the $N_2O_4 \rightleftharpoons 2NO_2$ equilibrium and to predict what they would observe and make an evaluation of their observations. (The increase in colour density on pressing the syringe is only transitory and often difficult to see as a new position of equilibrium is reached.)

- More able learners who finish early could also help struggling learners undertake the calculations when their own calculations have been checked.

Common learners misconceptions

- Some learners who have not had much experience with practical work have a limited idea of the capacity of common laboratory apparatus such as gas syringes or have difficulties relating the amounts of reagents used to the volume of gas collected. A demonstration of the capacity and limitations of various pieces of laboratory apparatus before the investigation is carried out can be useful.

- When drawing chemical apparatus, many learners leave gaps or draw lines across tubes where there should be a free flow of gas in the mistaken idea that as long as the basic apparatus is present, is does not matter how it is drawn or if it is labelled or not.

- Many learners will not realise that in demonstrating the effect of temperature on the $N_2O_4 \rightleftharpoons 2NO_2$ equilibrium, they are looking for a change in colour not a change in volume.

Method

1. A suitable apparatus set-up (see Figure 13.3) and method chosen for collecting the N_2O_4.

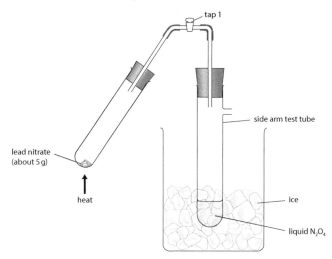

Figure 13.3

2. **a** Set up the apparatus in a fume cupboard with tap 1 open.

   b Heat the test tube containing about 20 g of lead nitrate **gently**.

   c Stop heating when a few cm³ of dinitrogen tetroxide have been collected.

   d Close tap 1 and keep the side arm test tube in the ice bath, then close the side arm.

3. A suitable apparatus set-up (see Figure 13.4) and method chosen for collecting the NO_2.

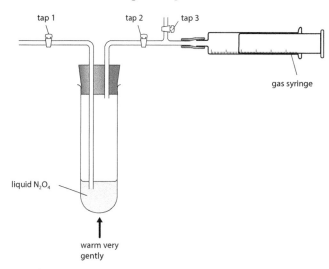

Figure 13.4

4. **a** Set up the apparatus in a fume cupboard. Keep tap 1 closed, tap 2 open and tap 3 open.

   b Warm the side arm test tube gently (not with a Bunsen).

   c Allow some gas to escape to expel air from the apparatus.

   d Fill the gas syringe with nitrogen dioxide (tap 3 closed).

   e Expel the gases from the gas syringe to remove contaminated air.

   f Close tap 3 and fill the gas syringe to about 60–80 cm³ with nitrogen dioxide.

5. A suitable method for demonstrating the effect of temperature on the $N_2O_4 \rightleftharpoons 2NO_2$ equilibrium.

   a Take two large beakers (e.g. 2 dm³). Fill one with water at 5 °C and the other with warm water (use a kettle) at 50 °C. The water levels should be high enough for most of the barrel of the syringe to be immersed.

   b Take two syringes containing the same volume of nitrogen dioxide. Put one in the cold water and one in the warm water.

   c Allow each syringe to stay in the cold or hot water until they have equilibrated.

d Compare the colour of the nitrogen dioxide in the two syringes.

- Learners should predict that the colour should be darker brown in the warmer water because the forward reaction is endothermic. The increase in temperature shifts the reaction in the endothermic direction which produces more brown nitrogen dioxide.

Sample results

In this investigation, learners are asked to select suitable values for the volume of gas to be collected. The rest of the results follow on from this. The volume suggested is 80 cm^3 for a 100 cm^3 gas syringe.

Answers to the workbook questions (using the sample results)

a 80 cm^3 NO$_2$

b mol = $\dfrac{80}{24000}$ = 3.3 × 10^{-3} mol

c Two mol NO$_2$ produced from one mol Pb(NO$_3$)$_2$

molar mass of Pb(NO$_3$)$_2$ = 331.2 so mass = 0.5 × 3.3 × 10^{-3} × 331.2 = 0.55 g

d Oxygen is given off in the reaction. Air is present in the empty tubes and can also come for the atmosphere through the open side arm.

e Bubble it through concentrated aqueous sodium hydroxide or pass it over moist calcium oxide.

f Air contains water vapour, which will dissolve nitrogen dioxide and interfere with the experiment; the air will dilute the colour of the equilibrium mixture and so make any changes less obvious.

g Nitrogen dioxide will dissolve in the water.

h It will be effective because there is a definite difference in colour; it may not be effective because the change in colour might not be very great.

Interpretation may be difficult because:

i the volume of the gas mixture also changes with temperature, a larger volume diluting the colour

ii the two syringes might take different times to equilibrate, so the colours are not being compared at the same time

iii if only one syringe is being used it is difficult to compare the colours unless photographs are taken.

Practical investigation 13.5: Data analysis
Investigation into equilibrium entropy and enthalpy change

Skills focus

The following skill areas are developed and practised (see the skills grids at the front of this guide for codes):

PI   Selecting information (c)
HI   Manipulating data (b, c)
DA   Identifying and using calculations (b)
CP   Drawing conclusions (b)
EI   Identifying problems with the data (a, c)

Duration

The data analysis and evaluation questions will take about 30 minutes depending on how much mathematical help is needed.

Preparing for the investigation

- Learners should have a basic understanding of entropy, enthalpy changes and should be able to substitute numerical values into mathematical expressions which are new to them.

- Learners should be able to deal with natural logarithms. Learners were introduced to these in the coursebook, Chapter 20 Electrochemistry.

Chapter 13: Further aspects of equilibria

Carrying out the investigation

- The information needed for this investigation is given in the workbook.

- Learners are asked to analyse data on the relationship between the equilibrium constant and the enthalpy change of reaction. They then draw a graph from which the enthalpy change can be determined.

  Some learners may need help with using their calculators to calculate the values of $\ln K_p$ in terms of both keying in numbers such as 1.21×10^{-4} and in using the "ln" button.

- Others may need help in drawing the graph because there are negative numbers involved.

- Many learners may need help with rearranging mathematical expressions which are unfamiliar to them.

  Learners who finish early could be asked to redraw the graph to extrapolate it to cut the y-axis when $\frac{1}{T}$ is zero. They could then determine the entropy change of the reaction from the value of the intercept which is $\frac{\Delta S^\circ}{R}$.

Common learners misconceptions

- Learners often forget to convert joules (per kelvin per mole) in the gas constant to kilojoules.

Answers to the workbook questions (using the sample results)

a See Table 13.3

Temperature /K	$\dfrac{1}{\text{Temperature}}$ / K^{-1}	K_p	$\ln K_p$
1800	5.556×10^{-4}	1.23×10^{-4}	-9.00
2000	5.000×10^{-4}	4.12×10^{-4}	-7.79
2200	4.545×10^{-4}	11.10×10^{-4}	-6.80
2400	4.167×10^{-4}	39.27×10^{-4}	-5.77
2600	3.846×10^{-4}	50.43×10^{-4}	-5.29
2800	3.571×10^{-4}	81.52×10^{-4}	-4.81

Table 13.3

b See Figure 13.5

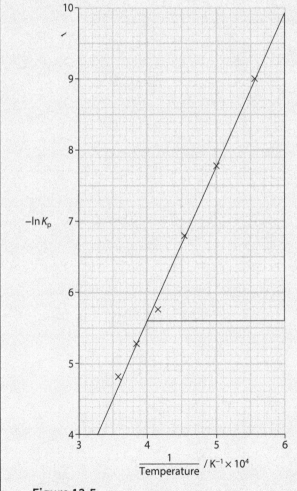

Figure 13.5

c   Example showing rise/run on graph (e.g. from 4×10^{-4} to 6×10^{-4} on the x-axis and corresponding figures on the y-axis)

$$\frac{10 - 5.55}{(6.0 - 4.0) \times 10^{-4}} = \frac{4.45}{2 \times 10^{-4}} = 22250$$

The gradient is positive so ΔH must be negative.

d ΔH = gradient × 8.314 = $-184\,987$ J mol^{-1} = -185 kJ mol^{-1} (to three significant figures)

e   There are no anomalous points. The points at ln K_p -5.77 and -4.81 do not deviate significantly from the best-fit line.

f   The points would be bunched at the top right of the graph and any slight errors in drawing the line would be magnified by extending this line further. To reduce this error, further values of K_p need to be obtained at a range of higher temperatures (e.g. 3000 K or 4000 K). Some learners may also point out that the data at very high temperatures may be subject to greater errors.

Chapter 14:
Reaction kinetics

Chapter outline

This chapter relates to Chapter 22: Reaction kinetics in the coursebook.

In this chapter, learners will complete investigations on:

- 14.1 Kinetics of the reaction between propanone and iodine
- 14.2 Data analysis investigation into rate of decomposition of an organic compound
- 14.3 Planning investigation into determination of the order of a reaction
- 14.4 Planning investigation into the effect of temperature on rate of reaction

Practical investigation 14.1:
Kinetics of the reaction between propanone and iodine

Skills focus

The following skill areas are developed and practised (see the skills grids at the front of this guide for codes):

PI	Selecting information (c)
	Defining the problem under investigation (a)
COI	Methods used (c)
	Carrying out the experiment (b)
HI	Collecting and displaying data (b)
	Manipulating data (a, b)
DA	Identifying trends and patterns (a)
	Identifying and using calculations (b)
CP	Drawing conclusions (b, c, d)
EI	Identifying problems with the procedure (a, b)
	Making a judgement on the conclusions (a, b)

Duration

- The practical work will take 30 minutes (or 1 hour if the experiment is repeated).
- The data analysis and evaluation will take 15 minutes.

Preparing for the investigation

- Learners should have some experience with an initial rate method for calculating rate of reaction.
- They should be able to analyse data from initial rate experiments in order to determine the order of a reaction.

Equipment

Each learner or group will need:

- 100 cm^3 conical flask
- stop clock or stopwatch
- test tube or boiling tube
- two 10 cm^3 measuring cylinders
- 20 cm^3 or 50 cm^3 measuring cylinder
- white tile or piece of white paper
- three sticky labels

Access to:

- 2.0 mol dm^{-3} propanone (50 cm^3)
- 2.0 mol dm^{-3} hydrochloric acid (100 cm^3)

- 0.01 mol dm⁻³ aqueous iodine (20 cm³)
- distilled water
- burette for dispensing distilled water

Additional notes and advice

To make 1 dm³ of aqueous iodine, dissolve 6 g of potassium iodide in 200 cm³ of distilled water, then add 2.54 g iodine (the solid is harmful) and dissolve. Make up to 1 dm³ with distilled water. If insufficient measuring cylinders are available, they could be shared amongst the groups.

Starch solution has not been included in the list of chemicals above because of evaluation question, h. Teachers, however, may wish to include a 1% starch solution and ask learners to add a few drops of this when the colour of the iodine becomes very faint.

Safety considerations

- Propanone is highly flammable and an irritant. There should be no naked flames in the laboratory. The propanone used in this experiment is diluted with water so presents less of a problem.
- Dilute iodine solution is low hazard.

Carrying out the investigation

- Learners are asked to analyse the kinetics of the reaction between propanone and iodine in the presence of hydrochloric acid.
- The major problem is cross-contamination of the measuring cylinders. This can be reduced if each measuring cylinder is labelled. The flasks need to be washed and dried between experiments. Ideally each group should have four flasks, one for each experiment.
- Some learners may need help in determining when the colour of the iodine has disappeared. Others may need help in understanding why this method is valid for all orders of reaction (question **c**) or in understanding how to use their results to determine the order of reaction (question **d**).
- Learners could be asked to find out more about the mechanism of this reaction or similar reactions using books or the internet.
- Learners who finish early could also help struggling learners analyse their data.

Common learner misconceptions

- The only misconceptions may arise from learners confusing time and rate.

Sample results

These give an idea of some results the learners may obtain in the investigation.

	Experiment A	Experiment B	Experiment C	Experiment D
Time for colour of iodine to disappear /s	67	283	308	131
Relative rate of reaction In cm³ I_2 /s	0.037	0.018	0.016	0.038

Table 14.1

Answers to the workbook questions (using the sample results)

a The concentration of iodine in Experiment A is half that of the other experiments since half the volume was used; 2.5 cm³ was used in Experiment A and 5 cm³ in the others.

b Just after the start of the reaction, the concentration of iodine appears to decrease at a constant rate with time. It is only when the concentration of iodine is low (below that which an obvious colour can be seen with the naked eye) that the differences in the slope of the curve become obvious.

c Comparing Experiments A and D (volume of propanone and HCl the same), doubling the concentration of I_2 has no effect on the rate. So it is zero order with respect to I_2. Comparing Experiments B and D (volume of HCl and I_2 the same), doubling the concentration of propanone doubles the rate. So it is first order with respect to propanone. Comparing Experiments C and D (volume of propanone and I_2 the same), doubling the concentration of HCl doubles the rate. So it is first order with respect to HCl. The overall order is second order.

Chapter 14: Reaction kinetics

> **d** Rate = $k[CH_3COCH_3][HCl]$ or Rate = $k[CH_3COCH_3][H^+]$
>
> **e** The hydrogen ions in the hydrochloric acid are involved in the rate-determining step of the reaction; the mechanism involves hydrogen ions in the slowest step.
>
> **f** Sufficient water is added to make the total volume of the reaction mixture the same so that the concentrations can be easily compared.
>
> **g** Measuring cylinders are inaccurate for small volumes: it would be better to use a burette for dispensing the solutions. The accuracy of the measuring cylinder may be to only the nearest 1 cm³ for the 50 cm³ cylinder but a burette is accurate to the nearest 0.05 cm³.
>
> **h** Iodine may be left in the bottom of the test tube when poured into the reaction mixture: they could find out how much iodine was left by adding water and doing thiosulfate titration. There could be difficulty in interpreting when the iodine is no longer visible: they could use a colorimeter to see when a particular depth of colour is obtained or add a few drops of starch solution when the colour of the iodine is very faint to sharpen the end-point. The temperature is not controlled: they could carry out the reaction in a thermostatically controlled water bath.

Practical investigation 14.2: Data analysis

Investigation into rate of decomposition of an organic compound

Extension investigation

Skills focus

The following skill areas are developed and practised (see the skills grids at the front of this guide for codes):

PI	Selecting information (c)
CHI	Collecting and displaying data (b) Manipulating data (b, c)
DA	Identifying trends and patterns (a, c)
EI	Identifying problems with the procedure (a) Identifying problems with the data (c) Making a judgement on the conclusions (a, b)

Duration

- The data analysis will take about 40 minutes and the evaluation less than ten minutes.

Preparing for the investigation

- Learners should be aware of the graphical methods for distinguishing a first-order reaction from a zero-order or second-order reaction. Learners should also know how to use the log button on their calculators and be able to plot points on a continuous graph which includes both positive and negative numbers.

- Some learners may worry about the unusual method of following the rate of this reaction and the introduction of an unfamiliar mathematical relationship.

- Learners should be reassured that all they have to do is follow the instructions. They could be told beforehand that this is a method that is different from those that they have come across before and a that if a reaction is first order, a plot of $\log_{10}(r_{final} - r_{at\ time\ t})$ against time will be a straight line with a downward gradient (r could be any measurable quantity such as volume or electrical conductivity).

- Some learners may get confused by the negative numbers obtained when the reaction is nearly complete.

- Learners who are more familiar with mathematics could calculate the value of the gradient and then the rate constant, k.

- Learners who finish early could also help struggling learners either analyse their data or help them plot their graphs.

- Learners study the hydroxide ion catalysed decomposition of an organic compound **Z** (4-hydroxy-4-methylpentan-2-one).

- As the reaction proceeds there is a small change in the volume of the reaction mixture, which is measured using a dilatometer. Learners analyse the results and plot graphs.

Common learner misconceptions

- Some learners who have not had much experience of mathematics may not distinguish between $\log_{10}$ and ln and so press the incorrect button on their calculators.

Answers to the workbook questions (using the sample results)

a/b See Table 14.2

Reaction using 0.20 mol dm⁻³ sodium hydroxide				Reaction using 0.05 mol dm⁻³ sodium hydroxide			
Time / min	Dilatometer reading / cm	$r_f - r_t$ / cm	$\log_{10}(r_f - r_t)$	Time / min	Dilatometer reading / cm	$r_f - r_t$ / cm	$\log_{10}(r_f - r_t)$
0	0.2	4.0	0.60	0	0.3	6.0	0.78
2	0.9	3.3	0.52	5	0.8	5.5	0.74
4	1.5	2.7	0.43	10	1.8	4.5	0.65
6	2.0	2.2	0.34	15	2.7	3.6	0.56
8	2.4	1.8	0.26	20	2.8	3.5	0.54
10	2.7	1.5	0.18	25	3.5	2.8	0.45
12	3.0	1.2	0.08	30	3.7	2.6	0.42
14	3.2	1.0	0	35	4.2	2.1	0.32
16	3.4	0.8	−0.10	40	4.3	2.0	0.30
18	3.6	0.6	−0.22	45	4.4	1.9	0.28
38	4.2	0		85	6.3	0	
40	4.2	0		90	6.3	0	

Table 14.2

c/d See Figure 14.1

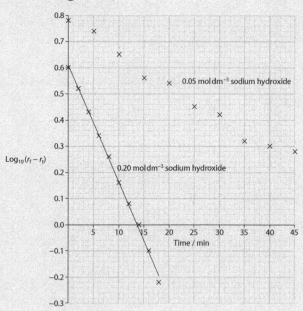

Figure 14.1

e If you plot $(r_f - r_t)$ against time, a curve is obtained from which you have to calculate the rate by taking tangents to the curve. There is a greater probability of error in drawing tangents than in drawing a line for $\log_{10}(r_f - r_t)$ which is a straight best-fit line.

f Z is in excess so its concentration does not effectively change during the reaction.

g Since the reaction is first order, the line for 0.05 mol dm⁻³ sodium hydroxide should also be a straight line of negative gradient. The data for this concentration becomes more variable towards the end of the experiment, where it appears to curve. So it is difficult to draw a best-fit line. This could be dealt with either by taking readings every 2 min to get more data or repeating the experiment. Learners could also comment on the fact that the data becomes more unreliable as the rate of reaction decreases when the reaction nears completion.

h A change is temperature can affect the volume of the glass bulb and the width of the capillary tube because of (relative) thermal expansion. Because of thermal expansion, the volume of the reaction mixture may also change. Since the change in length/volume of the mixture in the capillary tube is small, a small change in the volume of the bulb / width of the capillary tube will have a significant effect on the results.

i The readings were taken to the nearest mm / tenth of a cm. It is difficult to read values to the nearest 0.1 mm unless a magnifier is used. The difference between successive readings of the length/height of the liquid in the capillary tube is small, so there is likely to be a large percentage error in reading the values correctly. Small pieces of dirt or grease in the very fine capillary tube could affect the readings.

Practical investigation 14.3: Planning
Investigation into determination of the order of a reaction

Introduction
This is an extended investigation where students plan and carry out an experiment to determine the order of reaction when calcium carbonate reacts with hydrochloric acid.

Skills focus
The following skill areas are developed and practised (see the skills grids at the front of this guide for codes):

PI	Selecting information (a, b, c)
	Defining the problem under investigation (a, b, c, d)
	Control experiments and identification of variables (a, b, c)
	Considering hazards (a, b)
COI	Methods used (a, b, c, d)
	Carrying out the experiment (a, b, c, d)
HI	Collecting and displaying data (a, b)
	Manipulating data (a, b, c)
DA	Identifying trends and patterns (a, b, c)
CP	Drawing conclusions (b, c, d, e)
EI	Identifying problems with the procedure (a, c, d)
	Identifying problems with the data (c)
	Making a judgement on the conclusions (a)

Duration
- This is an extended investigation. The planning work will take 40 minutes and it is suggested that this is done for homework.
- The practical work will take at least 30 minutes (or 120 minutes if the candidates suggest an initial rate method using different starting concentrations of hydrochloric acid).
- The analysis and evaluation questions will take 40 minutes.

Preparing for the investigation
- Learners should have some experience of a variety of methods used for determining rate of reaction.
- Learners should also be aware of how to determine rate from either graphs of volume of gas produced over time (measuring the gradient at different points on the graph) or from the initial rate method (measuring the initial rate using different initial concentrations of acid).
- It is useful if the learners carry out initial experiments to determine the amount and size of the particles and concentration of hydrochloric acid to use.

Carrying out the investigation
- Learners should realise the marble chips should be in excess. Learners should show this by calculation (e.g. 10 cm^3 of 1.0 mol dm^{-3} HCl = 0.01 mol). So, according to the mole ratio in the equation, at least 0.005 mol CaCO_3 is required (e.g. more than 0.5 g). So use 4 g of $CaCO_3$. A calculation of the volume of gas expected should also be done for Method A (e.g. maximum volume of gas required = $100 \text{ cm}^3 = \frac{100}{24000} \text{ mol} = 4.17 \times 10^{-3} \text{ mol } CO_2$). So, according to the mole ratio in the equation, $8.34 \times 10^{-3} \text{ mol}$ HCl is required.

- Learners are asked to plan and carry out one or more experiments to determine the order of reaction with respect to hydrochloric acid when this acid reacts with marble chips.

- The most common problems arise from the need to read the gas syringe and take timings every 10 or 20 s. For Method B, the number of readings can be reduced by taking them every 20 s after one minute.

- A problem with both methods is that carbon dioxide is soluble in aqueous solutions and some will dissolve in the acid. This problem can be reduced if the timing is not started too early. Since this is a first-order reaction, the rate of change is constant.

- Each learner plan should be approved before they carry out the experiment to ensure that the method is viable and that some attempt has been made to control the temperature.

- Some learners may need help at the start to understand how to calculate the masses, volumes and concentrations of the reagents required.

- Many learners will benefit from conducting a preliminary experiment to help select appropriate masses, volumes, concentrations and intervals of timing. Others may need help when setting up the apparatus.

- Some learners may need help in handling the data in an appropriate manner and plotting relevant graphs.

- Learners could be asked to devise more than one graphical method of determining the order of reaction or compare the results from the two different methods.

- Learners who finish early could also help struggling learners either finish their experiment or plot suitable graphs.

Common learner misconceptions

- Some learners may consider rate being proportional to time rather than inversely proportional to time.

Equipment

The equipment chosen by each learner or group will vary as will their labelled apparatus set-up. The following is a suitable list for each learner or group:

- boiling tube with side-arm or small side-arm flask (Method A)
- 100 cm^3 gas syringe
- stopper to fit test tube or flask
- rubber connecting tubing
- stopclock or stopwatch
- 10 cm^3 or 20 cm^3 measuring cylinder,
- 100 cm^3 conical flask (Method B)
- clamp stand and two clamps (Method A)

Access to:

- 1.0 mol dm^{-3} hydrochloric acid
- 6.0 mol dm^{-3} hydrochloric acid (if requested)
- marble chips
- distilled water
- cotton wool
- top-pan balance to weigh to the nearest 0.01 g
- weighing boats

The marble chips should be prepared by washing in very dilute hydrochloric acid to remove marble dust. They are then washed with distilled water and air dried.

Safety considerations

- Eye protection should be worn.
- If learners suggest using concentrations of hydrochloric acid of concentration 6.0 mol dm^{-3}, they should realise that it is corrosive. Concentrations of hydrochloric acid between 2.0 mol dm^{-3} and 6.0 mol dm^{-3} are an irritant.

Method

There are two alternative methods.

Method A

1. Connect a side-arm boiling tube or side-arm flask to a gas syringe. Clamp the syringe and boiling tube. The boiling tube or flask should be stoppered.
2. Weigh out about 10 g marble chips (each chip about 1.5-2 g).
3. Place 8 cm^3 of 1.0 mol dm^{-3} hydrochloric acid in a side-arm flask or side-arm boiling tube.
4. Add the marble chips to the acid and start the stopclock.
5. Read the volume of the gas syringe every 20 s until the volume remains constant.

Additional notes and advice

- Some learners may wish to conduct a variant on this method by doing several experiments starting with different initial concentrations of hydrochloric acid, perhaps concentrations higher than 2.0 mol dm^{-3}

Method B

1. Place 10 cm^3 of 1.0 mol dm^{-3} hydrochloric acid in a conical flask.
2. Put a plug of cotton wool in the mouth of the flask.
3. Place the flask and its contents on a top-pan balance.
4. Weigh out 10 g marble chips (each chip about 1.5-2 g).
5. Add the marble chips to the flask, record the balance reading and start the stopclock.
6. Take the balance reading every 20 s until the mass remains constant.

For both methods:

- Independent variable = time
- Dependent variable = volume of carbon dioxide or loss in mass of reaction mixture
- Other variables that need to be controlled include temperature, because the rate of reaction increases with increase in temperature. This could be controlled by carrying out the experiment in a thermostatically controlled water bath. This may not be necessary because the reaction is complete within 3 min. If several repeats of the experiment are done, this becomes more important. The mass of marble chips and their approximate surface area also need to be the same because the greater the surface area, the more particles are exposed to react.

Chapter 14: Reaction kinetics

Sample results

Tables 14.3 and 14.4 provide sample results the learners may obtain in the investigation.

Method A

Time /s	0	20	40	60	80	100	120	140	160	180	200	220
Volume of CO_2 / cm^3 (run 1)	0	24	46	59	69	76	81	85	88	90	91	91
Volume of CO_2 / cm^3 (run 2)	0	26	45	61	72	80	84	87	89	91	92	92

Table 14.3

Method B

Time /s	0	20	40	60	80	100	120	140	160	180	200	220
Mass / g (run 1)	83.57	83.42	83.34	83.24	83.18	83.15	83.13	83.12	83.11	83.10	83.10	83.10
Mass /g (run 2)	82.32	82.18	82.07	82.01	81.93	81.89	81.87	81.85	81.84	81.83	81.82	81.82

Table 14.4

Answers to the workbook questions (using the sample results)

a There are several methods of analysing the data:

i Plot a graph of volume of carbon dioxide against time. Take gradients at different points on the graph (see the coursebook) and plot a graph of rate against concentration. This should give a line showing a directly proportional relationship between rate and concentration.

ii Plot a graph of (final volume of carbon dioxide – volume of carbon dioxide at time t) against time (or final mass loss – mass loss at time t) against time. This is because the (final volume – volume at time t) is proportional to the concentration of the hydrochloric acid at time t. If the graph has a constant half-life, the reaction is first order. If the half-life increases as time increases, it is second or third order (see Chapter 22 in the coursebook).

iii Some learners may do several experiments and calculate the initial rate using different concentrations of hydrochloric acid. The graph of rate against time should show a directly proportional relationship between rate and concentration.

iv Figure 14.2 is a graph of (final volume of carbon dioxide – volume of carbon dioxide at time t) against time.

b The rate of reaction increases with increase in concentration of hydrochloric acid. The reaction is first order because:

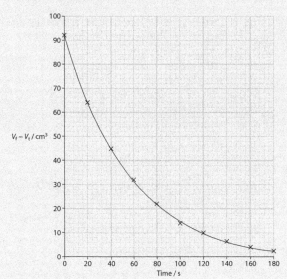

Figure 14.2

i A graph of rate against concentration of acid shows that as the concentration of hydrochloric acid increases, the rate of reaction increases in direct proportion.

ii A graph of (final volume of carbon dioxide – volume of carbon dioxide at time t) against time or (final mass loss – mass loss at time t) against time shows a half-life which is constant.

Cambridge International AS & A Level Chemistry

> **c** Points all fall on a smooth curve so data is reliable or points do not all fall on a smooth curve (points are scattered) so data is not reliable. Repeated experiment shows similar curve (points of similar values) so data is reliable or repeated experiment shows scatter points with dissimilar values so data is not reliable.
>
> **d** Difficulty in reading the syringe or balance and the stopclock at exactly the same time leads to random errors in either timing or taking the reading. The syringe plunger is in constant motion at the beginning of the experiment so there may be errors in the volume of carbon dioxide produced. Carbon dioxide dissolves in the reaction mixture so the volume (or mass) recorded is too low at the start. Small volumes and concentrations of hydrochloric acid were used so there would have been errors in the amount of hydrochloric acid present (learners might suggest the use of a burette to dispense hydrochloric acid rather than a measuring cylinder). Loss of carbon dioxide from the boiling tube and flask immediately after the addition of the marble chips and before the connection to the gas syringe makes the final volume too low. Marble chips are not of constant size and surface area. If insufficient marble chips are added, the surface area may decrease during the experiment. Temperature is not controlled and the rate of reaction depends also on temperature.

Practical investigation 14.4: Planning
Investigation into the effect of temperature on rate of reaction

Skills focus

The following skill areas are developed and practised (see the skills grids at the front of this guide for codes):

PI   Selecting information (b, c)
      Defining the problem under investigation (a, c, d, e)
      Control experiments and identification of variables (a, b, c)
      Considering hazards (a, b)

COI Methods used (a, b, c, d)
      Carrying out the experiment (a, b, c, d)

HI   Collecting and displaying data (a, b, c)
      Manipulating data (a, b, c)

DA   Identifying trends and patterns (a, b, c)

CP   Drawing conclusions (b, c, e)
      Making predictions (a, b)

EI   Identifying problems with the procedure (a, d)
      Identifying problems with the data (c)
      Making a judgement on the conclusions (a, b)

Duration

- This is an extended investigation. The planning work will take 30 minutes, but this could be done for homework.
- The practical work will take at least 60 minutes (or 120 minutes if the results are repeated). The analysis and evaluation questions will take 30 minutes.

Preparing for the investigation

- Learners should have some experience of a variety of methods used for determining rate of reaction as well as understanding the concept of rate.
- Before starting the experiment, learners should be aware how to warm a solution gently to a particular temperature.
- They should also be aware of the importance of using separate measuring cylinders for different solutions to avoid cross contamination.

Carrying out the investigation

- Learners are asked to plan and carry out a series of experiments to show how the rate of reaction of sodium thiosulfate with hydrochloric acid changes with temperature.
- The commonest problems arise from contamination of the solutions. The measuring cylinders used should be labelled, one for hydrochloric acid and one for sodium thiosulfate. A third measuring cylinder for water may also be required. If the measuring cylinders are confused, sulfur will start to precipitate and invalidate the experiment.
- If the second method is used, the test tubes X and Y must be used for the same solutions for each run of the experiment. The flask in the first method or tube X in the second method must be thoroughly rinsed with water between runs to prevent excess acid reacting with the fresh sample of sodium thiosulfate added.
- Temperature control in the first method is very poor because the contents of the flask start to cool. An average temperature can be taken but this is still unsatisfactory. At all temperatures, there will be considerable differences in the rate of precipitation of sulfur as the temperature falls.

Chapter 14: Reaction kinetics

- ⚙ Each learner's plan should be approved before they carry out the experiment to ensure that the method is viable and that some attempt has been made to control the temperature.

- Some learners may need help at the start to understand how the method will work. Others may need help when setting up the apparatus, in keeping the apparatus free from cross-contamination and in keeping the apparatus clean.

- Some learners may need help adding a last column to their table (rate in terms of the reciprocal of time taken for the cross/dot to be obscured).

- ⚙ Learners could be asked to plot a suitable graph to determine the activation energy of the reaction (ln(rate) against $\frac{1}{T}$ in kelvin).

- Learners who finish early could also help struggling learners to either finish their experiment or plot their graphs.

Common learner misconceptions

- Some learners may consider rate being proportional to time rather than inversely proportional to time.

- Others who just devise the basic experiment may not take note of the drop in temperature as the reaction mixture cools and just focus on the initial temperature.

Equipment

The equipment chosen by each learner or group will vary as will their labelled apparatus set-up. The following is a suitable list for each learner or group:

- eye protection
- 400 cm³ or 500 cm³ beaker
- two boiling tubes
- Bunsen burner, tripod and gauze
- 0–110°C thermometer
- two 10 cm³ measuring cylinders
- 50 cm³ measuring cylinder (optional)
- sticky labels (to label measuring cylinders)
- stopclock or stopwatch
- clamp stand and two clamps
- glass stirring rod
- alternative to beaker and two boiling tubes = 250 cm³ conical flask

Access to:

- 0.1 mol dm⁻³ sodium thiosulfate
- 0.5 mol dm⁻³ hydrochloric acid
- water
- black marker pen
- kettle (optional)

Safety considerations

- Eye protection should be worn.

- Learners should be aware that the sulfur dioxide produced in the reaction is toxic. Unless larger volumes of sodium thiosulfate are used, the amount of sulfur dioxide produced is small as long as the room is well ventilated.

- Learners who are asthmatic should carry out the experiment in a fume cupboard.

- If learners suggest using concentrations of hydrochloric acid at 2.0 mol dm⁻³ or slightly above, they should be aware that it is an irritant. The reaction mixture should be disposed of by pouring it down a sink with cold running water, preferably in a fume cupboard. It should be noted that, the higher the temperature, the more sulfur dioxide is released. It is not advisable to heat the reaction mixture to temperatures above 65 °C because of this and because of the difficulty with dealing with hot glassware.

Method

The simplest method is as follows.

Method A

1. Place 10 cm³ aqueous sodium thiosulfate and 40 cm³ water in a conical flask.

2. Warm the sodium thiosulfate solution in the flask to the temperature required (e.g. 25 °C) using a Bunsen burner.

3. Place the conical flask over a sheet of paper with a black cross draw on it.

4. Add 5 cm³ 2 mol dm⁻³ hydrochloric acid to the flask and swirl the flask to mix the contents

5. Immediately start the stopclock and record the temperature.

6. Look down through the reaction mixture from above. When you can no longer see the cross, stop the clock and record the final temperature.

7. Pour the reaction mixture down the sink; rinse and dry the flask.

8 Repeat the method five or six times using different temperatures.

A better method is as follows.

Method B

1 Add 10 cm³ aqueous sodium thiosulfate to one boiling tube. Label this tube X.

2 Add 10 cm³ 0.5 mol dm⁻³ hydrochloric acid to a second boiling tube. Label this tube Y.

3 Place a piece of paper marked with a black ink spot on the outside of a large beaker.

4 Fill the 500 cm³ beaker with warm water from a kettle (or heat the water using a Bunsen burner) to just above the temperature required (e.g. 28 °C if you want to conduct the experiment at 25 °C).

5 Clamp the boiling tubes in the water so that the solutions in them are completely immersed and so that you can see the black ink spot by looking through tube X.

6 Allow both tubes to come to the same temperature as the water (e.g. 25 °C) then quickly pour the contents of tube Y (acid) into tube X (thiosulfate) and stir with the thermometer.

7 Immediately start the stopclock and record the temperature.

8 Keep the temperature constant by adding more hot water and stirring if necessary.

9 When you can no longer see the black ink spot by looking through tube X, stop the clock and record the final temperature.

10 Pour the reaction mixture down the sink; rinse and dry the tubes.

11 Repeat the method five or six times using different temperatures. If there is time, repeat the whole experiment at the same temperatures.

For both methods:

- Independent variable = temperature

- Dependent variable = time taken for the cross/spot to be obscured

- Other variables that need to be controlled are: volumes and concentrations of hydrochloric acid and sodium thiosulfate; distance of surface of the liquid from the cross (or thickness of the boiling tube / same boiling tube used).

Additional notes and advice

- Method B is more appropriate for learners at this level as there is an attempt to control the temperature. Some learners may suggest using a thermostatically controlled water bath. These could be used if available. Some learners may suggest including temperatures below room temperature. This could be done with the use of ice.

Sample results

- Predicting results: Learners should realise that as the temperature increases there is an increased number of collisions with energy greater or equal to the activation energy. Weaker learners may concentrate on the fact that particle collisions are more energetic at higher temperatures or simply that the particles have more kinetic energy. Learners should therefore predict that a graph of the reciprocal of time against temperature should be an upward sloping line (either linear or curve).

- Effectiveness of the procedure: Candidates should comment on the fact that, despite the problems with temperature control, the experiment is likely to be effective because the temperatures selected are far enough apart for the timings to be significantly different. In addition, they could comment on the use of the same depth of solutions. Other learners could suggest that the procedure may not be effective because of the opposite reasons.

Method A

Experiment	Temperature of mixture at start / °C	Temperature of mixture at end / °C	Average temperature / °C	Time taken for cross to disappear / s	$\dfrac{1}{\text{Time}}$ / s⁻¹
1	20	21	20.5	89	0.011
2	33	31	32	82	0.012
3	48	45	47.5	28	0.357
4	60	56	58	8	0.125
5	69	65	67	7	0.143

Table 14.5

Method B

Experiment	Temperature / °C	Time taken for cross to disappear (run 1) / s	Time taken for cross to disappear (run 2) / s	Average time taken for cross to disappear / s	$\dfrac{1}{\text{Time}}$ / s^{-1}
1	21.0	91	79	85	0.012
2	32.5	67	57	62	0.016
3	40.5	26	34	30	0.031
4	53.0	14	20	17	0.059
5	61.5	10	6	8	0.125

Table 14.6

Answers to the workbook questions (using the sample results)

a See Figure 14.3 for a graph of the results using Method B.

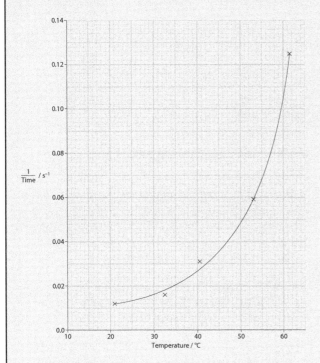

Figure 14.3

The plot of the results for Method A shows a general increase in rate as temperature rises but it is difficult to draw an acceptable curve. Learners should comment on the variability of this data. For Method B the curve is far better and the data could be used to calculate a value for the activation energy.

b The rate of reaction increases with increase in temperature. If a suitable curve is obtained, learners should comment on the shape of the curve; for example, the rate increases more and more for each 10 °C temperature rise or the increase appears to be exponential. Reference could also be made to time being inversely proportional to rate.

c The experiment agrees closely with predictions because the rate increases as the temperature increases (or the experiment doesn't agree entirely with predictions because in some cases, the increase in temperature has little effect on the rate).

d It is difficult to control the temperature because the temperature of the reaction mixture decreases when the flask is removed from the heat; continued heating would increase the temperature of the reaction mixture. Accuracy of the thermometer is not to the nearest 0.1 °C. Inaccuracy in timing at high temperatures may be considerable because the reaction only takes a few seconds to complete and there may be a lag time in starting and stopping the clock. It is difficult to determine exactly when the black cross/spot is completely obscured.

e The data is insufficient because: there were not enough repeats at different temperatures (because of significant differences in readings); it was difficult to replicate readings at the same temperature; results were not consistent enough at high temperatures. The data is sufficient, because three or more replicate readings were taken at a range of (at least five) temperatures and these were consistent.

f Use a thermostatically controlled water bath or add hot water to keep the temperature constant when the temperature of the reaction mixture starts to fall. Use a thermometer reading to 0.1 °C. Use a light sensor connected to a computer to monitor the intensity of the precipitate with time. Bring both the hydrochloric acid and the sodium thiosulfate to the same temperature before mixing.

Chapter 15:
Transition elements

Chapter outline

This chapter relates to Chapter 24: Transition elements in the coursebook.

In this chapter, learners will complete investigations on:

- 15.1 Planning investigation into copper content of copper ore
- 15.2 Data analysis investigation of iron tablets
- 15.3 Data analysis investigation into formula of a complex ion
- 15.4 Planning investigation into reaction of copper with potassium dichromate(VI)

Practical investigation 15.1: Planning
Investigation into copper content of copper ore

Skills focus

The following skill areas are developed and practised (see the skills grids at the front of this guide for codes):

PI	Selecting information (c)
	Defining the problem under investigation (a, c, d)
	Considering hazards (a, b)
COI	Methods used (a, b)
	Carrying out the experiment (c)
HI	Manipulating data (b, c)
DA	Identifying and using calculations (a)
CP	Drawing conclusions (a)
	Making predictions (a, b)
EI	Identifying problems with the procedure (a)
	Making a judgement on the conclusions (b)

Duration

- The planning will take 30 minutes (or 90 minutes if practical work is included); the analysis and evaluation questions will take 10 minutes.
- The experiment does not always give good results and you may consider this as a planning exercise only.
- The preparation of the solutions of different concentrations is the most time-consuming step.

If doing the experiment, you may consider preparing the solutions of copper(II) sulfate of different concentrations beforehand.

Preparing for the investigation

- Learners should have some experience of making up a less concentrated solution from a more concentrated solution. They should also revise the relationship between moles, concentrations and volume of solution.

Carrying out the investigation

- This investigation asks learners to plan a simple experiment to attempt to determine the mass of copper ions in malachite by treating the ore with sulfuric acid and then comparing the depth of colour of the copper(II) sulfate solution with range of pre-prepared copper(II) sulfate solutions of known concentration.

- The different dilutions of copper(II) sulfate solutions should be labelled carefully with the appropriate concentrations so that they are not confused.

- The test tubes should be of the same thickness to avoid errors in matching the colour intensity. If the colour intensity of the solutions are low, it is better to make a comparison by looking through the same length of liquid column from the top of the tubes rather than through the tubes.

- Some learners may need help with understanding how dilutions are made.

- Other learners may need help in describing how the results are processed.

Chapter 15: Transition elements

⚙ Learners who finish early could be asked to find out about how colorimetry works and how and why it is useful in the analysis of transition element ions.

Common learner misconceptions

- Some learners may forget to scale up the values so that all the copper is accounted for, for example, in taking only a fraction of the solute from the filtration for analysis.

Equipment

Each learner or group may suggest:

- plastic gloves
- mortar and pestle / hammer
- spatula
- 200 cm³ beaker
- 100 cm³ measuring cylinder
- filter funnel and filter paper
- 250 cm³ conical flask
- stopper for flask
- eight test tubes and test-tube rack

Access to:

- top-pan balance and weighing boat
- burette containing 1.0 mol dm⁻³ copper(II) sulfate
- burette for distilled water
- 2.0 mol dm⁻³ sulfuric acid (60 cm³)
- 1.0 mol dm⁻³ aqueous copper(II) sulfate (20 cm³)
- two beakers and funnels for filling each burette
- powdered malachite or substitute

Additional notes:

- This investigation is a planning exercise. It is not intended that the learners should do the experiment themselves but the equipment list above is suitable if teachers wish to demonstrate the procedure.
- It is not intended that the learners should break up the malachite themselves, since this would be too dangerous because of pieces of flying rock. Some may suggest using a mortar and pestle for grinding the small pieces.
- A suggested substitute for powdered malachite is a mixture of fine sand (80%) and basic copper carbonate (20%).

Safety considerations

- Gloves and eye protection should be worn when dispensing the sulfuric acid since it is corrosive. The sulfuric acid must be added slowly to the malachite to stop acid splash as the carbon dioxide / gas escapes.
- Grind the malachite carefully to minimise raising of dust and wear a mask while grinding so that no dust is breathed in.

Method

Part 1

1 Break up the malachite into small pieces (with a hammer) and grind with a mortar and pestle.

2 Add 50 cm³ 2.0 mol dm⁻³ sulfuric acid a little at a time.

3 When there are no more bubbles in the reaction mixture after adding the sulfuric acid, filter the mixture into a flask.

4 Wash the residue with a little distilled water.

5 Pour the filtrate into a measuring cylinder, rinse the flask with distilled water and add the rinsings to the measuring cylinder.

6 Add distilled water to the measuring cylinder until the total volume is 100 cm³; record the volume of copper(II) sulfate in the measuring cylinder.

Part 2

1 Prepare a range of dilutions of the 1.0 mol dm⁻³ aqueous copper(II) sulfate so that the total volume is the same, for example, 2 cm³ 1.0 mol dm⁻³ copper(II) sulfate + 8 cm³ water, 4 cm³ 1.0 mol dm⁻³ copper(II) sulfate + 6 cm³ water.

2 Place a known volume of the solution of unknown concentration in the measuring cylinder into a test tube. Label this tube U.

3 Place the same volume of each of the diluted solutions of copper(II) sulfate into separate labelled test tubes.

4 Compare the colour of tube U with these test tubes to see which matches the best. Dilute the known solutions further if necessary.

5 Learners could also suggest gradually diluting a known volume of the 1.0 mol dm⁻³ aqueous copper(II) sulfate solution with distilled water and shaking. When the solution has the same colour intensity as solution U, the volume of water added is recorded.

Sample results

- Some learners might find it easier to answer the data analysis question by using specific values (e.g. mass of malachite taken = 20 g; volume of copper(II) sulfate filtrate = 100 cm^3; volume of U, taken for comparison = 10 cm^3; dilution of 1.0 mol dm^{-3} aqueous copper(II) sulfate taken 4 cm^3 copper(II) sulfate + 6 cm^3 water).

Answers to the workbook questions (using the sample results)

a Concentration of diluted copper(II) sulfate =
$1.0 \times \dfrac{4}{10}$ mol dm^{-3} = 0.4 mol dm^{-3}

Moles of copper ions in 10 cm^3 solution
= $0.4 \times \dfrac{10}{1000}$ mol = 4×10^{-3} mol

Mass of copper ions in 10 cm^3 solution
= $4 \times 10^{-3} \times 63.5$ = 0.254 g

Mass of copper ions in 100 cm^3 solution (obtained from all the ore) = 2.54 g

b A burette because of its greater precision; a burette can be read to the nearest 0.05 cm^3 (between the 0.1 cm^3 graduation marks) but a 20 cm^3 measuring cylinder can only read to the nearest 0.2 cm^3.

c Make a further range of dilutions in between those which have already been suggested.

d Use test tubes of the same width and make the height of the liquid column the same so that the depth of solution viewed is the same.

e There is difficulty in comparing the colours accurately since there is a gradation in tone. The thickness of the walls or of the tubes may be different. The tubes reflect light so it is difficult to compare the colours. The ore is hard and cannot be ground up into a fine enough powder, so not all the copper ions are extracted. There may be loss of solid material when bubbles of carbon dioxide escape on adding sulfuric acid.

Practical investigation 15.2: Data analysis
Investigation of iron tablets

Skills focus

The following skill areas are developed and practised (see the skills grids at the front of this guide for codes):

PI   Selecting information (b)

HI   Collecting and displaying data (b)
            Manipulating data (a, b)

DA   Identifying trends and patterns (c)
            Identifying and using calculations (a, b)

CP   Drawing conclusions (b)

EI   Identifying problems with the procedure (a)
            Identifying problems with the data (b, c)
            Making a judgement on the conclusions (a)

Duration

- The data analysis will take 20 minutes and the evaluation will take 20 minutes.
- If time is available, learners could carry out the experiment themselves. This will add an extra 50 minutes to the investigation.

Preparing for the investigation

- Learners should already have experience of titrations and redox reactions. Ensure that they are aware of the use of acidified potassium manganate(VII) as an oxidising agent and how to use the stoichiometric equation to deduce the number of reacting moles.

- Learners may not be familiar with the ease of oxidation of Fe^{2+} to Fe^{3+} by atmospheric oxygen, so some mention of this may be needed by demonstrating the 'browning' of iron(II) hydroxide when left in air for some time.

Carrying out the investigation

- This investigation is an extension of a standard titration technique in which learners are asked to calculate the percentage of iron(II) ions in an iron tablet, following a titration of the Fe^{2+} in the iron tablet with acidified potassium manganate(VII).

- In some tablets, glucose may be present, which is also a reducing agent. Glucose reacts with acidified potassium manganate(VII) much more slowly than Fe^{2+} and so should not influence the results significantly.

- If insufficient acid is present, a brown or red colouration may be seen in the titration flask. If this occurs, more sulfuric acid should be added.

Chapter 15: Transition elements

- The end-point of the titration is not always obvious so there may be some variation in the results. Shaking the flask carefully, and for at least 5 s following the addition of each drop of potassium manganate(VII), is essential.

  🧑 Some learners may need help with some of the practical procedures. For example, in part **i** they may not know how to transfer all the paste to the flask by washing out the mortar.

- Other learners may need help with the titration calculations.

- If learners undertake the experiment, they may need help in determining the end-point of the titration.

  ⚙ Learners could be asked to find out about other redox titrations involving potassium manganate(VII).

Common learner misconceptions

- When doing the calculations, some learners may not take into account the stoichiometry of the reaction.

Equipment

Each learner or group will need:

- burette and clamp
- 100 cm³ volumetric flask
- funnel
- 100 cm³ titration flask
- white tile
- 10 cm³ volumetric pipette
- pipette filler

Access to:

- mortar and pestle
- iron tablets (from local pharmacy)
- 1.0 mol dm⁻³ sulfuric acid
- 0.0050 mol dm⁻³ standardised potassium manganate(VII)
- distilled water to rinse titration flask

Safety considerations

- Learners should wear eye protection throughout the experiment.
- 1.0 mol dm⁻³ sulfuric acid is irritant to eyes and skin.
- Solid potassium manganate(VII) is harmful and oxidising but 0.0050 mol dm⁻³ potassium manganate(VII) is low hazard.
- Iron tablets are low hazard.

Answers to the workbook questions (using the sample results)

a See Table 15.1

	Run 1	Run 2	Run 3	Run 4
Initial burette reading / cm³	0.10	5.15	9.65	14.85
Final burette reading / cm³	5.25	9.65	14.85	19.70
Titre / cm³	5.15	4.50	5.20	4.85

Table 15.1

b The products have very pale colour so appear almost colourless. When all the Fe^{2+} ions have reacted, the potassium manganate(VII) is in excess and the purple colour becomes visible. This is the end-point of the titration.

c It is likely to be too high because the end-point is unknown so you do not know at what point you have to add the potassium manganate(VII) drop by drop.

d The titre for Run 3 is significantly different from the titres for Runs 2 and 4 (about 17% difference). Runs 2 and 4 are slightly more consistent but still have about 7% difference. The titration should be repeated with the same solution until more consistent results are obtained.

e mol MnO_4^-(aq) = $\frac{4.5}{1000} \times 0.0050 = 2.25 \times 10^{-5}$ mol

f moles Fe^{2+} in 10 cm³ = $2.25 \times 10^{-5} \times 5 = 1.125 \times 10^{-4}$ mol

moles Fe^{2+} in 100 cm³ = $2.25 \times 10^{-4} \times 10 = 1.125 \times 10^{-3}$ mol

$1.125 \times 10^{-3} \times 55.8 = 0.063$ g (to 2 significant figures).

g % by mass = $\frac{0.063}{0.58} \times 100 = 10.9\%$

h It supplies the large amount of hydrogen ions required by the stoichiometric equation and it suppresses the hydrolysis of Fe^{2+} ions.

i Wash the paste through a funnel into the volumetric flask. Wash the mortar out several times with small volumes of sulfuric acid and pour the washings through the funnel until all the paste has been removed from the mortar.

j Stopper the flask and turn it upside down several times.

k Pour the solution from the volumetric flask into a beaker then take the $10\,cm^3$ sample using a $10\,cm^3$ volumetric pipette with pipette filler.

l It will make it more difficult to observe the correct colour change at the end-point of the titration.

m The end-point of the titration may be difficult to observe because very low concentrations of (excess) potassium manganate(VII) may not be visible and the pale yellow colour of the Fe^{3+} ions formed may make it difficult to determine the end-point. Not all the Fe^{2+} ions are released from the iron tablets by grinding and washing since it is mixed with insoluble material and binding material. Some of the Fe^{2+} ions may be oxidised on exposure to air or hydrolysed in water and so the titres will be too low.

Practical Investigation 15.3: Data analysis
Investigation into formula of a complex ion

Skills focus

The following skill areas are developed and practised (see the skills grids at the front of this guide for codes):

PI   Selecting information (c)
      Defining the problem under investigation (e)
      Control experiments and identification of variables (b)

HI   Collecting and displaying data (b)
      Manipulating data (b, c)

DA   Identifying trends and patterns (a, c, d)
      Identifying and using calculations (b)

CP   Drawing conclusions (b, c)

EI   Identifying problems with the procedure (b)
      Identifying problems with the data (a, c)
      Making a judgement on the conclusions (a, b)

Duration

- The data analysis will take 20 minutes and the evaluation will take 20 minutes.

- If time is available, teachers could demonstrate the experiment with the learners taking readings of the colorimeter. This may take an extra 15 minutes.

Preparing for the investigation

- Learners should have some idea of the use of a colorimeter in following the progress of a reaction, but they may need help in understanding how a colorimeter works and with the idea of choosing an appropriate colour filter which will maximise the absorption by the coloured solution. The details required are those given in the question.

- Learners should have an understanding of the term transmittance before they undertake the investigation (e.g. copper(II) sulfate solution appears blue because blue light is transmitted and other colours are absorbed).

Carrying out the investigation

- Some learners may need help with logarithms.

- This investigation uses the idea of continuous variation of solution concentration which was introduced in investigation 15.1. Learners attempt to find the ratio in which Ni^{2+} ions combine with $EDTA^{4-}$ ions to form a complex ion by colorimetry. They determine the ratio of these ions when the colour of the solution is most intense. It should be noted that this method is only suitable where only one species is formed.

- Readings of the colorimeter may not be stable, especially if a needle-type instrument is used.

- It is essential that stray light does not enter the colorimeter cell, otherwise the correct transmittance will not be recorded.

Chapter 15: Transition elements

- ⚙ Some learners may need help in understanding the shape of the graph and that the greater the concentration, the less light is transmitted.

- Other learners may need help in answering question **d**, where learners could be given a hint that they could make a range of solutions of known concentration.

- Some might need help in question **e**, which depends on a correct answer to question **d**.

- ⚙ Learners who finish early could use books or the internet to find out more details about spectroscopy in the visible or ultraviolet regions of the spectrum.

Common learner misconceptions

- Some learners may confuse the transmittance of light with the absorption of light.

Equipment

Each learner or group will need:

- 11 boiling tubes with stoppers (for making solutions)
- test-tube racks

Access to:

- colorimeter with cell
- 0.050 mol dm^{-3} nickel(II) sulfate (60 cm^3)
- 0.050 mol dm^{-3} EDTA (60 cm^3)
- distilled water

Safety considerations

- Solid and 0.050 mol dm^{-3} nickel(II) sulfate are harmful.
- Solid EDTA is irritant but 0.050 mol dm^{-3} EDTA is low hazard.

Answers to the workbook questions (using the sample results)

a See Table 15.2

Volume of 0.05 mol dm^{-3} Ni^{2+} ions / cm^3	Volume of 0.05 mol dm^{-3} EDTA^{4-} ions / cm^3	Colorimeter reading / % transmittance	$\frac{I_o}{I}$	$\log_{10} \frac{I_o}{I}$
0	10	100	1.00	0
1	9	74	1.35	0.13
2	8	55	1.82	0.26
3	7	50	2.00	0.30
4	6	23	4.37	0.64
5	5	22	4.55	0.66
6	4	26	3.85	0.59
7	3	35	2.86	0.46
8	2	39	2.56	0.41
9	1	63	1.59	0.20
10	0	91	1.10	0.04

Table 15.2

b See Figure 15.1

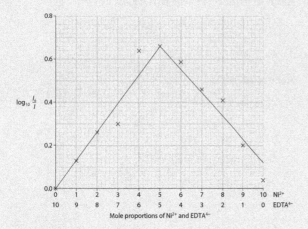

Figure 15.1

c 5:5 Ni^{2+}:EDTA^{4-}. The highest concentration of complex is present when the intensity of colour is greatest, i.e. when the absorption is greatest. At other concentrations, the complex is more dilute.

d Make a range of solutions of known concentrations. Take the meter readings of each solution. Plot the readings against the concentrations. A straight line going through the origin will indicate proportionality but a slight curve to the line will indicate that it is not (always) proportional.

e A measurement which is proportional to concentration is all that is needed.

In this case it is $\log_{10} \frac{I_o}{I}$, which is proportional to concentration.

f The mole proportions of Ni^{2+}:EDTA^{4-} at 4:6 and 5:5 are very similar so the formula could be either Ni$_2$(EDTA)$_3$ or Ni(EDTA). Four of the first six points fall on the same line. The other two (3:7 and 4:6 ratios) may be anomalous and so can be ignored. The last five points show considerable deviation from the line so it is difficult to take these into account. The last five points are more consistent with the ratio being 5:5 rather than 4:6.

g Increase the number of different volumes take; for example, Ni^{2+} 4.5 cm^3:EDTA^{4-} 5.5 cm^3. Repeat the experiment using the same volumes until consistent results are obtained.

h The water is a control experiment to account for any absorption of light by the water, so the amount of light transmitted by the water and tube is kept constant (since the concentrations are very low).

i The width of the test tubes may not be the same so the width of the liquid will be different; wider tubes will absorb more light or transmit less light. The walls of the test tubes may be different so more light may be absorbed by thicker test tubes. There may be difference in refractive index of the glass leading to different amounts of light reflected and transmitted.

Practical Investigation 15.4: Planning
Investigation into reaction of copper with potassium dichromate(VI)

Skills focus

The following skill areas are developed and practised (see the skills grids at the front of this guide for codes):

PI   Selecting information (c)
      Defining the problem under investigation (a, c, d, e)
      Control experiments and identification of variables (b, c)
      Considering hazards (a, b)

COI Methods used (a, b, d)
      Carrying out the experiment (b)

HI   Collecting and displaying data (b)

DA Identifying trends and patterns (a, b, d)

EI   Making a judgement on the conclusions (a)

Duration

- The planning will take 40 minutes; the analysis and evaluation questions will take 15 minutes.
- This is a planning exercise with an extension for data analysis and evaluation.
- The experiment must not be attempted because of the high toxicity of potassium dichromate.

Preparing for the investigation

- This investigation extends the idea of colorimetry using a single filter (investigation 15.3) to the use of spectrometry in the whole of the visible region of

Chapter 15: Transition elements

the spectrum. Learners should have some idea that a visible spectrometer is able to scan a solution at a variety of wavelengths.

- Learners should have an understanding of the term transmittance before they undertake the investigation; for example, Cu^{2+} ions appear blue because they transmit blue light and other colours are absorbed, whereas Cr^{3+} ions appear greenish because they transmit green light (and some red) and other colours are absorbed.

Carrying out the investigation

- Learners are asked to plan an experiment to show the changes in the transmission of light at different wavelengths when copper reacts with acidified potassium dichromate to form Cu^{2+} ions and Cr^{3+} ions

  Some learners may need help in understanding the spectra. Cu^{2+} ions appear blue because they transmit blue light and other colours are absorbed, whereas Cr^{3+} ions appear greenish because they transmit green light (and some red) and other colours are absorbed.

- The colour of the potassium dichromate ions has not been given; learners made need some help in deducing that the orange dichromate will transmit light in the yellow and red parts of the visible spectrum with green and blue being absorbed.

- Some learners may need help with drawing the spectrum for the reaction when it is about half complete.

  Learners who finish early could deduce the order of reaction with respect to dichromate ions using the following information about the % transmittance at 610 nm (reflecting the decrease in concentration of dichromate ions). 0 min: 98%; 10 min: 79%; 20 min: 73%; 30 min: 63%; 40 min: 46%; 50 min: 41%; 60 min: 38%; 80 min: 35%; 100 min: 33%; 150 min: 31%.

Common learner misconceptions

- Some learners may confuse the transmittance of light with the absorption of light.

Equipment

Each learner or group may suggest:

- Eye protection and plastic gloves
- Beaker or conical flask (for reaction)
- Volumetric flasks (for making solutions)
- Two burettes or graduated pipettes
- Stirring rod or magnetic stirrer
- Filter funnel (for burette)
- Graduated pipette and pipette filler (for removing samples)
- Stopwatch or stopclock

Access to:

- Top-pan balance, weighing boat and spatulas
- Spectrometer and cells for spectrometer
- Distilled water (for making solutions)
- 50 cm^3 or 100 cm^3 measuring cylinders
- Sandpaper (for cleaning copper foil)

Suitable volumes and concentrations for each component of the reaction mixture.

- Learners should calculate the number of moles of copper and potassium dichromate to show that the copper is in excess of 3 mol copper to 1 mol dichromate and that the dichromate is at a concentration less than 0.05 mol dm^{-3}.

- The sulfuric acid should also be in at least 14 times more concentrated than the potassium dichromate.

- For example, to make 100 cm^3 of reaction mixture without the copper:

  1. If concentration of potassium dichromate = 0.02 mol dm^{-3} take a number of moles of H_2SO_4 more than 14 × moles potassium dichromate, i.e. 20 × 0.02 = 0.4 mol dm^{-3}.

  2. Therefore, dilute the 2.00 mol dm^{-3} H_2SO_4 solution provided by 5. Take 100 cm^3 of this solution.

  3. 0.02 mol dm^{-3} potassium dichromate = 294.2 × 0.02 × 0.1 g = 0.588 g in 100 cm^3

  4. Dissolve 0.588 g potassium dichromate in 100 cm^3 of 0.4 mol dm^{-3} sulfuric acid.

  5. The mass of copper should be > 3 × 0.02 × 0.1 mol = > 6 × 10^{-3} mol = > 0.38 g.

Method

Each learner or group may suggest:

1. Place 100 cm^3 of the mixture of potassium dichromate and sulfuric acid in a beaker.
2. Place on magnetic stirrer.
3. Add 1 g clean copper foil, stir the solution and start the stopclock.

4 Keep the solution stirred.

5 After 5 minutes / stated time interval, remove a fixed volume, for example, 5 cm³ from the reaction mixture using a graduated pipette and place in the spectrometer a cell. Use cell containing water as a control.

6 Record the spectrum.

7 Remove further samples at regular intervals and record the spectrum again until there is no change in the spectrum.

Safety considerations

This is a planning exercise only. The investigation should not be carried out practically for safety reasons.

Each learner or group may suggest:

- Weigh the potassium dichromate using gloves and do not raise dust (of potassium dichromate).
- Wear a mask to prevent dust from reaching your nose or mouth.
- Wear eye protection to stop dust reaching yours eyes.
- Use eye protection or mask and cover arms when using the solution of potassium dichromate.
- Use gloves and eye-protection when diluting the sulfuric acid.

Answers to the workbook questions (using the sample results)

a/b See Figure 15.2

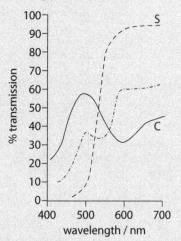

Figure 15.2

b The peak of the combination of Cu^{2+} ions and Cr^{3+} ions around a wavelength of 400–500 nm, where blue and green light is transmitted, will be lower than that when the reaction is complete because fewer Cu^{2+} ions and Cr^{3+} ions are present. The transmission 550–700 nm, where yellow and red light is transmitted, will be lower than at the start of the experiment because dichromate ions are being removed from the reaction mixture.

c For one particular wavelength (e.g. 600 nm), determine the % transmission of the dichromate ions at different time intervals, or for one particular wavelength (e.g. 500 nm), determine the % transmission of the chromium + copper ions. For dichromate ions, plot a graph of %transmission against time and analyse the shape of the curve using the half-life method. For the chromium + copper ions, a graph of (final transmittance − transmittance at time t) is plotted against time.

d The water is a control experiment to account for any absorption of light by the water; so the amount of light transmitted by the water in the solution is kept constant (since the concentrations are very low).

e Any two of the following are valid:

  i Temperature: rate of reaction changes markedly if temperature is increased or decreased.

  ii Same width of solvent used / same cell used: The wider/longer the pathway the light has to travel, the more the light is absorbed by the solution and the less light is transmitted through the solution.

  iii Same cell used: Cells made of different materials may absorb/refract light by different amounts.

  iv Same amount of stirring used: Different rates of stirring may cause changes in the collision frequency of the copper with the dichromate ions; there may be local differences in concentration of dichromate ions.

f The reaction is not continuing because no solid copper is present. The volume of the sample of the reaction mixture does not matter as long as the cell is full and as long as the width/length of the solution in the cell is constant.

g A higher concentration may not allow the transmittance of much light. The initial changes in transmittance of dichromate ions or later changes in transmittance of chromium and copper ions cannot be easily determined.

Chapter 16:
More about organic chemistry

Chapter outline

This chapter relates to Chapters 25: Benzene and its compounds, Chapter 26: Carboxylic acids and their derivatives, Chapter 27: Organic nitrogen compounds and Chapter 28: Polymerisation in the coursebook.

In this chapter learners will complete investigations on:

- 16.1 Planning investigation into making an azo dye
- 16.2 Data analysis investigation into acylation of a nucleic acid
- 16.3 Planning investigation into nitration of benzene

Practical Investigation 16.1: Planning
Investigation into making an azo dye

Skills focus

The following skill areas are developed and practised (see the skills grids at the front of this guide for codes):

PI   Selecting information (b, c)
        Defining the problem under investigation (a, c, d, e)
        Control experiments and identification of variables (b)
        Considering hazards (a, b)

COI   Methods used (a, b, d)

Duration

- This is a planning exercise only. Planning will take about 50 minutes. This includes doing the relevant calculations in the method section.
- The evaluation questions will take 20 minutes.

Preparing for the investigation

- Learners should have some experience of the chemistry of amines and phenols.
- Before starting the investigation, learners should know how to calculate volume from mass and density as well as perform simple mole calculations to determine which substances are in excess in a given reaction mixture.

Carrying out the investigation

- Learners plan an experiment to prepare a sample of the azo dye formed when phenylamine undergoes diazotisation and the product is then coupled with 2-naphthol to form the azo dye, benzene-azo-2-naphthol.

- The investigation must not be done practically because of the hazardous nature of the chemicals involved. An alternative demonstration experiment could be conducted to form a different azo-dye using ethyl-4-aminobenzoate (irritant and flammable) in place of phenylamine using the same method as outlined below, as long as the safety considerations mentioned below are in place.

  Some learners may need help with working out the precise details of how to calculate the amount of each substance used.

- Other learners may not realise that they need to prepare all the solutions at the start of the experiment and carry out Stage 2 as quickly as possible after Stage 1 in order to minimise the decomposition of the intermediate diazonium salt.

- Some learners may need a hint that they need to be able to separate the azo dye from the rest of the reaction mixture and the method of doing this using a Buchner flask and filter paper connected to a water pump.

  More able learners who finish early could use books or the internet to investigate the range of colours of azo dyes in terms of the type of amine and phenol used. They could also be asked to construct chemical equations for some of these reactions.

- Others could be asked to make an apparatus list, stating the sizes of the glassware used.

Common learner misconceptions

- Some learners may have little idea of the amount of reactants to use to keep potential hazards to a minimum. Some may suggest using quantities in the range of 0.2–1 mole, which is excessive.

Equipment

- Learners should select a small amount of starting materials to keep potential hazards to a minimum.

- It may be useful to give learners a value of 0.01 mol of phenylamine to use. They can then calculate the moles of 2-naphthol and sodium nitrite required for these to be in excess.

  If 0.01 mol phenylamine is used (M_r = 93), the mass of phenylamine = 0.93 g and the volume = 0.93 ÷ 1.02 = 9.1 cm^3.

  Mass of 2-naphthol (M_r = 144) > 0.01 mol, for example, 0.015 mol = 2.16 g (2 g is sufficient).

  Mass of sodium nitrite (M_r = 69) > 0.01 mol, for example, 0.015 mol = 1.04 g (1 g is sufficient).

  10 cm^3 of 2 mol dm^{-3} hydrochloric acid to dissolve the phenylamine.

  20 cm^3 of 2 mol dm^{-3} aqueous sodium hydroxide to dissolve the 2-naphthol. This should be in excess of the hydrochloric acid.

Safety considerations

- The experiment should be carried out in a fume cupboard and a mask and eye protection should be worn in addition to a lab coat.

- Gloves must be worn when dispensing the phenylamine, 2-naphthol and sodium nitrite.

- A pipette filler should be used when dispensing the phenylamine.

- Care must be taken not to raise dust when weighing out the sodium nitrite.

- There should be no naked flames in the laboratory.

Method

A suggested method is:

1 Use a graduated pipette with a pipette filler to put 0.9 cm^3 of phenylamine into a 100 cm^3 beaker then add about 10 cm^3 of 2 mol dm^{-3} hydrochloric acid and stir.

2 Place the beaker in an ice bath and let the temperature fall to about 5 °C.

3 Dissolve 1 g sodium nitrite in 10 cm^3 of ice-cold water.

4 In a separate beaker (250 cm^3), dissolve 2 g of 2-naphthol in 20 cm^3 of 2 mol dm^{-3} sodium hydroxide.

5 Step 1: Add the sodium nitrite solution a little at a time to the cold phenylamine solution and stir with a thermometer. Make sure that the temperature does not rise above 10 °C (add some ice to the reaction mixture if the temperature rises too rapidly).

6 Step 2: Leave the solution in Step 1 to react for a few minutes then pour it slowly into the solution containing 2-naphthol and sodium hydroxide. Stir after each addition.

7 Leave the solution to react then filter off the solid azo dye using a Buchner funnel and water pump (or filter funnel and filter paper).

8 Wash the residue with distilled water and leave to dry in the air.

Answers to the Workbook questions (using the sample results)

a So that all the phenylamine is reacted, and there is still enough nitrous acid present even if some decomposes.

b The heat evolved in the reaction may raise the temperature too quickly and cause the diazonium salt to decompose.

c The temperature is kept below 10 °C during the first stage of the experiment so that no decomposition occurs.

d Decomposition of nitrous acid / diazonium salt may take place because temperature control is not exact. Some azo dye may be left on the sides of the beaker / on the filter paper / on the sides of the Buchner funnel.

e Recrystallise by dissolving the azo dye in a minimum volume of warm solvent then cooling and filtering. Repeat as many times as necessary.

f Heat it gently using a melting point apparatus. Record the melting point. If the melting point is sharp and is not lower than published data, the sample is pure.

> **g** Prepare a control experiment by using the sodium nitrate and hydrochloric acid only without using the phenylamine.
>
> **h** They may get into the drainage system and be harmful (especially the phenylamine) to organisms in the water.

Practical investigation 16.2: Data analysis
Investigation into acylation of a nucleic acid

Extension investigation

Skills focus

The following skill areas are developed and practised (see the skills grids at the front of this guide for codes):

PI	Selecting information (c)
	Control experiments and identification of variables (b, c)
HI	Collecting and displaying data (b)
	Manipulating data (b, c)
DA	Identifying trends and patterns (a, b)
CP	Drawing conclusions (b)
EI	Identifying problems with the procedure (c)
	Making a judgement on the conclusions (a, b)

Duration

This is a data analysis exercise. The investigation will take about 50 minutes, including the evaluation questions.

Preparing for the investigation

- Learners should have some knowledge of the properties of amino acids and know that proteins and nucleic acids are natural polymers whose monomers are amino acids and nucleotides respectively. They should also have some knowledge of how enzymes work (see Chapter 9 in the coursebook).

- Before starting the investigation, learners should be introduced to the method of following a reaction by using a radioactively 'labelled' reactant and measuring the increase in radioactivity (in counts per minute) in the product as the reaction progresses.

Carrying out the investigation

- Learners are asked to analyse how pH affects the rate of incorporation of radioactive proline into tRNA in the presence of an enzyme extracted from a plant.

- Some learners may need help with the concept of an enzyme-catalysed reaction and with how the progress of a reaction can be monitored using a reactant which is radioactive.

- Others may need help with drawing a suitable line between the points on the graph.

- Some learners may need help at the start with the concept of using a control experiment to account for background radioactivity.

- Learners who are interested in Biology could be asked to find out more about protein synthesis.

- Other learners could be given some pH buffer calculations to do, for example, to work out how to make a buffer solution of a specific pH given relevant information about the acid and conjugate base.

Common learner misconceptions

- Many learners will be used to drawing a best-fit curve through the points on a graph and may think that they should do so here. However, the points do indicate that there is a definite change in gradient at about pH 7.

Answers to the Workbook questions (using the sample results)

a See Table 16.1

pH	Run 1 Counts / min	Run 1 Corrected counts / min	Run 2 Counts / min	Run 2 Corrected counts / min	Average Corrected counts / min
5.80	10	4	12	4	4
6.00	14	8	16	8	8
6.15	21	15	42	34	25

pH	Run 1 Counts / min	Run 1 Corrected counts / min	Run 2 Counts / min	Run 2 Corrected counts / min	Average Corrected counts / min
6.45	53	47	62	54	51
6.60	68	62	75	67	65
6.75	80	74	84	74	74
6.95	91	85	89	81	83
7.10	93	87	98	90	89
7.25	96	90	100	92	91
7.55	102	96	106	98	97
7.85	106	100	112	104	102
8.15	116	110	116	108	109
8.75	114	108	116	108	108
9.00	111	105	115	107	106
9.50	104	98	118	110	104
10.0	106	100	112	104	102

Table 16.1

b See Figure 16.1

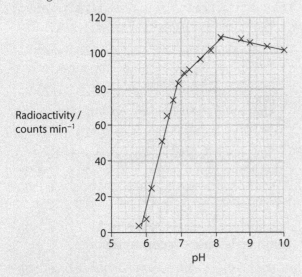

Figure 16.1

c Temperature. The rate of reaction increases rapidly as temperature increases.

d If there are significant readings of background radiation, these need to be taken into account by subtracting these values from the experimental values – the idea of a control experiment.

e At low pH values, the background radiation is a significant proportion of the total radioactivity; for example, for pH 5.8 the background radiation is greater than the radiation in the prolyl-tRNA. So large errors may arise. In addition, only the average background radiation has been measured so even a difference in 1 count / min will influence the results significantly. At higher pH values, the error is low because the radiation in the prolyl-tRNA is much greater than the background radiation.

f It has been assumed that the background radiation is constant, which it is unlikely to be. It also assumes that the rate of incorporation of radioactivity from the reactant (proline) to the product (prolyl-tRNA) is linear with time.

g To stop the reaction; to precipitate the tRNA and protein.

h Proline / amino acids are fairly soluble in aqueous solution so the radioactive proline can be separated from the precipitate of tRNA/protein by filtration.

i The pH increases rapidly from pH 6 to 7. From pH 7 to 8, the pH increases more slowly. Above pH 8 the pH decreases slowly. To show that there are three separate portions to the graph, more repeat readings should be taken between pH 6.5 and 8.

j So that any (small amounts of) acid or base produced in the reaction will not change the pH.

k Differences in the composition of the buffer solution or different amounts of maleic acid and tris may affect the enzyme/tRNA.

l There may be other impurities or enzymes present which will break down the enzyme/tRNA or alter the functioning of the enzyme.

Practical Investigation 16.3: Planning
Investigation into nitration of benzene

Skills focus

The following skill areas are developed and practised (see the skills grids at the front of this guide for codes):

PI   Selecting information (b, c)
Defining the problem under investigation (a, c, d)
Control experiments and identification of variables (b)
Considering hazards (a, b)

COI   Methods used (d)

Duration

This is a planning exercise only. Planning will take about 50 minutes. This includes doing the relevant calculations in the method section. The evaluation questions will take 20 minutes.

Preparing for the investigation

- Learners should have some experience of the chemistry of benzene.

- Before starting the investigation, learners should know how to calculate volume from mass and density as well as perform simple mole calculations to determine which substances are in excess in a given reaction mixture.

- Learners should understand the use of a separating funnel in separating liquids of differing densities.

Carrying out the investigation

- Learners plan an experiment to prepare a sample of nitrobenzene from benzene.

- On no account should the investigation be done practically because of the hazardous nature of the chemicals involved.

  🎧 Some learners may need help with working out the precise details of how to calculate the amount of each substance used.

- Other learners may not realise how to use the information given to help them prepare a plan for the preparation. Some may need to be given a hint how to separate the nitrobenzene from the reaction mixture (the densities should give a clue) and how to remove the excess acidity from the mixture of nitrobenzene, acid and water.

  🎧 More able learners who finish early could be use books or the internet to investigate how to prepare crystals of dinitrobenzene from nitrobenzene.

Common learner misconceptions

- Some learners may have little idea of the amount of reactants to use to keep potential hazards to a minimum. Some may suggest using quantities of acid in the range of one mole, which is excessive.

Equipment

- If 0.08 mol benzene is used (M_r = 78), the mass of benzene = 6.24 g and volume = 6.24 ÷ 0.88 = 7.1 cm³. So 7 cm³ is sufficient.

- For sulfuric acid to be in excess, we need 0.1 mol = 9.8 g. Volume = 9.8 ÷ 1.84 = 5.3 cm³ so 6–10 cm³ is sufficient.

- For nitric acid to be in excess, we need 0.1 mol = 6.3 g. Volume = 6.3 ÷ 1.51 = 4.2 cm³ so 6–10 cm³ is sufficient.

Method

1. Use a graduated pipette with a pipette filler to put 10 cm³ of concentrated sulfuric acid into a 100 cm³ distillation flask, then add about 10 cm³ of concentrated nitric acid from a graduated pipette a little at a time with stirring.

2. Add 1 cm³ portions of benzene to the mixture while stirring, making sure that the temperature does not rise above 50 °C. Continue until the 7 cm³ benzene has been added.

3. Connect the flask to a water condenser in the upright position for reflux.

4. Place the flask in a water bath set at 60 °C and when the temperature is reached, heat for 30 minutes.

5. After 30 minutes, let the flask cool, then pour the contents of the distillation flask into some cold water (50–100 cm³) in a beaker and stir.

6. Allow the layers (nitrobenzene and aqueous layer) to separate and remove most of the upper (aqueous) layer with a pipette (or decant off most of the upper layer).

7. Put the rest of the mixture into a separating funnel and add aqueous sodium carbonate (to react with the excess acids).

8 Stopper and shake the separating funnel and then allow the layers to separate. Then pour off the upper layer.

9 Repeat this process until no more bubbles are seen on addition of aqueous sodium carbonate.

10 Transfer the lower (nitrobenzene) layer to a small flask and add solid sodium sulfate (to remove water) and stir until the nitrobenzene is not cloudy.

11 Distil the nitrobenzene (simple distillation), collecting the fraction boiling at about 208–212 °C.

Safety considerations

- In their answers, learners should mention some of the safety aspects (e.g. the experiment should be carried out in a fume cupboard and a mask and eye protection should be worn in addition to a lab coat).
- Gloves must be worn when dispensing each liquid.
- A pipette filler should be used when dispensing the liquids.

Answers to the workbook questions (using the sample results)

a So that all the benzene is reacted.

b Corrosive vapours are more likely to escape. The heat evolved in the reaction may raise the temperature too quickly and cause acid to spit out of the beaker. This is less likely with a flask.

c The acids are added together slowly and the benzene is added to the acid slowly to minimise the loss of vapour produced by the heat produced during the reaction. The reaction mixture is heated under reflux, which reduces the loss of vapours. The reaction mixture is heated at a fairly low temperature to reduce the amount of vapour formed.

d Some nitrobenzene may be lost in decanting the water. Some remains locked up with the sodium sulfate used to dry it.

e Redistil it gently and record the temperature at which it distils. Use boiling point apparatus and record the boiling point. If the boiling point is sharp and is not higher than published data, the sample is pure.

Chapter 17:
Identifying organic compounds

Chapter outline

This chapter relates to Chapter 29: Analytical chemistry in the coursebook.

In this chapter learners will complete investigations on:

- 17.1 Data analysis investigation into extracting an amino acid from hair
- 17.2 Data analysis investigation into identification of a white crystalline solid
- 17.3 Data analysis investigation into preparation and identification of a colourless liquid

Practical investigation 17.1: Data analysis
Investigation into extracting an amino acid from hair

Skills focus

The following skill areas are developed and practised (see the skills grids at the front of this guide for codes):

PI	Selecting information (a, b, c)
	Defining the problem under investigation (d)
COI	Methods used (a, b)
HI	Collecting and displaying data (b)
DA	Identifying trends and patterns (a)
CP	Drawing conclusions (b, e)
EI	Making a judgement on the conclusions (a, b)

Duration

This is a data analysis exercise. The investigation will take about 50 minutes but up to 20 minutes extra time may need to be allowed for learners to access information about the Lassaigne test.

Preparing for the investigation

- Learners should have some knowledge of the properties of amino acids (especially that they are fairly soluble in water) and know that proteins are natural polymers. They should also know how to calculate R_f values from the results of paper chromatography.

- Either before, or during the investigation, learners should be introduced to the method of testing for sulfur and nitrogen – the Lassaigne test. Learners are asked to research this method for themselves. This could be done for homework.

Carrying out the investigation

- Learners answer questions about the preparation of a sample of amino acid A (cystine) from hair and then attempt to identify the amino acid by using chemical tests and chromatography.

  Some learners may need help with the questions on the practical procedure.

- Other learners may need help with clarifying the details of how the Lassaigne test is used to confirm the presence of sulfur and nitrogen in compound **A**.

  Learners who finish early could be given a mass spectrometer trace of cysteine to analyse (see Spectral database for organic compounds SDBS at sdbs.db.aist.go.jp). They could be given the parent ion peak at 240 and try to suggest the reason for the peaks at 64 (S–S), and 44 (COO).

Common learner misconceptions

- Some learners, when calculating R_f values, may take the measurement from the top of the spot rather than the middle.

Answers to Workbook questions (using the sample results)

a Any suitable organic solvent, for example, tetrachloromethane or hexane. The idea that grease/fats are non-polar so a non-polar solvent needed to dissolve it.

b See Figure 17.1

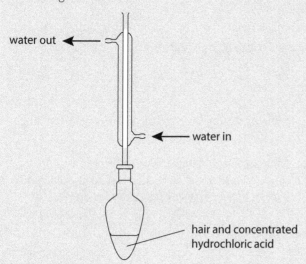

Figure 17.1

c To hydrolyse the proteins in the hair to amino acids; to minimise the loss of acidic vapours when heating.

d Add concentrated aqueous sodium hydroxide drop by drop until blue litmus paper just turns red (or until Universal Indicator just turns green or a pH meter reads 7). To adjust the solution to pH 5, add a weak acid or concentrated sodium ethanoate until pH 5 is reached.

e Filter off crystals. Wash the crystals with ethanol. Leave the crystals to dry in air, a drying oven, or a at low-temperature oven.

f Dissolve the crystals in a minimum volume of hot water. Filter immediately (using Buchner funnel and flask) into a flask with the water pump full on. Allow the solution in the flask to crystallise. Then filter off the crystals.

g Carboxylic acid.

h Sulfur is present (in the Lassaigne sodium fusion test, sulfur is reduced to sulfide ions which produce hydrogen sulfide when warmed with hydrochloric acid. Hydrogen sulfide turns lead ethanoate paper brownish-grey).

i Nitrogen is present (in the Lassaigne sodium fusion test, nitrogen forms cyanide ions which form specks of Prussian blue when treated with iron(II) sulfate and then acidified iron(III) chloride).

j A: $R_f = \frac{1.45}{5.0} = 0.29$ so most likely to be cystine

Contaminating amino acid $\frac{2.3}{5.0} = 0.46$ so most likely to be glycine

k No, arginine has an R_f which is not significantly different to cystine and histidine has an R_f which is not significantly different to glycine. Run a known sample of cystine on the same piece of chromatography paper at the same time as the sample and compare the distance travelled **or** use a different solvent which separates cysteine and arginine to a greater extent **or** use a longer piece of chromatography paper and run for a longer time.

Practical investigation 17.2: Data analysis
Investigation into identification of a white crystalline solid

Skills focus

The following skill areas are developed and practised (see the skills grids at the front of this guide for codes):

PI Selecting information (a, b, c)
    Defining the problem under investigation (d)

COI Methods used (a, b)

HI Collecting and displaying data (b)
    Manipulating data (a)

DA Identifying trends and patterns (a, b, d)

CP Drawing conclusions (b, c, d, e)

EI Making a judgement on the conclusions (a, b)

Duration

This is a data analysis exercise. The investigation will take between 35 and 50 minutes depending on the ability of the learner.

Preparing for the investigation

- Learners should have some knowledge of the properties of phenols and carboxylic acids and of the hydrolysis of esters.

Chapter 17: Identifying organic compounds

- Before the investigation, ensure that learners have some familiarity with mass spectroscopy and infrared spectroscopy.

Carrying out the investigation

- Learners answer questions which lead them to identify compound **X** (phenyl benzoate), compound **Y** (phenol) and compound **Z** (benzoic acid). They do this by considering some physical and chemical properties of these compounds and the mass spectrum and infrared spectrum of **Y**.

- Some learners may need help with the organic reactions involved. They could be given hints by referring them to specific pages in the coursebook.

- Other learners may need help with interpreting the infrared spectrum and in reaching a conclusion about the nature of the three compounds.

- Learners who finish early could be given a mass spectrometer and infrared spectra for phenyl benzoate to analyse.

- Others, once their work has been checked, could help struggling learners with the questions.

Common learner misconceptions

- Some learners may try to read too much into the fragmentation pattern of the mass spectrometer trace or try to identify more peaks in the infrared spectrum than necessary.

Answers to the Workbook questions (using the sample results)

a   Place a given mass of solid X (known number of moles) into a flask. Add excess aqueous sodium hydroxide. Fit the flask with a reflux condenser and heat the flask gently until all X dissolves.

b   It is very weakly acidic (because, although its pH is below 7, it is not acidic enough to release carbon dioxide from sodium carbonate). It has some non-polar nature (because it does not dissolve very well in water).

c   94 (parent ion peak)

d   Reaction with benzoyl chloride, C_6H_5COCl, to give $C_{13}H_{10}O_2$ suggests that Y has the formula C_6H_6O, which could be phenol or unsaturated alcohol / aldehyde. White precipitate with bromine water suggests phenol. This fits with M_r being 94 and the fact that phenol is acidic but not acidic enough to be a carboxylic acid.

e   Broad band at around either side of $3250\,cm^{-1}$ suggests hydrogen-bonded phenol (or alcohol). Spike at around $1500\,cm^{-1}$ suggests aromatic compound.

f   Use a tube containing reaction mixture with bung and delivery tube leading into limewater or a test tube containing reaction mixture with a glass rod suspended above with a drop of limewater attached or collect gas in a Pasteur pipette then bubble through limewater. Limewater turns cloudy/milky if the gas is carbon dioxide.

g   Z: $C_{13}H_{10}O_2 + H_2O$ (from hydrolysis) − $C_6H_5OH = C_7H_6O_2$. This compound is an acid since CO_2 is given off with carbonate, so is likely to be C_6H_5COOH (benzoic acid). So X is the ester, phenyl benzoate $C_6H_5COO\,C_6H_5$

Practical investigation 17.3: Data analysis
Investigation into preparation and identification of a colourless liquid

Skills focus

The following skill areas are developed and practised (see the skills grids at the front of this guide for codes):

PI	Selecting information (a, b, c)
HI	Collecting and displaying data (b) Manipulating data (a, b, c)
DA	Identifying trends and patterns (a, b, d)
CP	Drawing conclusions (b, c, d, e)

Duration

This is a data analysis exercise. The investigation will take 35 minutes.

Preparing for the investigation

- Learners should have some knowledge of the properties of aldehydes and ketones and the iodoform test.
- Before the investigation, ensure that learners have some familiarity with mass spectroscopy and infrared spectroscopy.

Carrying out the investigation

- Learners answer questions about the preparation of compound **R** (ethanal) and then identify it using chemical tests and spectral data.

    Some learners may need help with the organic reactions involved. They could be given hints by referring them to specific pages in the coursebook.

- Other learners may need help with interpreting the infrared spectrum and in reaching a conclusion about the nature of **R**.

     Learners who finish early could be given a mass spectra, infrared spectra or NMR spectra of ethanal and ethanol to analyse.

Common learner misconceptions

- Some learners may try to read too much into the fragmentation pattern of the mass spectrometer trace or try to identify more peaks in the infrared spectrum than necessary.

Answers to the Workbook questions (using the sample results)

a See Figure 17.2

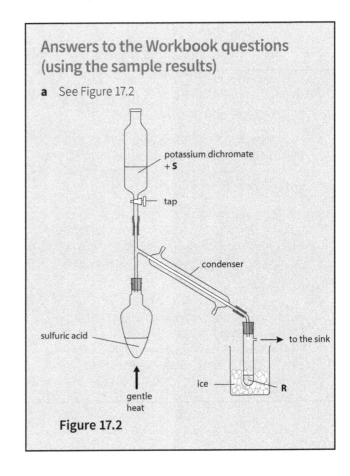

Figure 17.2

b i Dissolve excess potassium dichromate, for example 5 g, in the minimum volume of water and add about 5 cm³ of liquid S. Shake to dissolve.

ii Set up the apparatus with sulfuric acid in the flask and the mixture of potassium dichromate and S in the tap funnel.

iii Warm the acid then turn off the Bunsen burner.

iv Open the tap funnel and add the mixture of potassium dichromate and S to the sulfuric acid slowly.

v Distil R, keeping the mixture in the flask gently warmed so that the temperature does not rise above 70 °C.

c R is an aldehyde; R contains a group which reduces Cu^{2+} ions.

d R contains the CH_3CO group.

e $M_r = 44$ (parent ion peak)

f $29 = CHO^+$ group, $15 = CH_3^+$ group

g CH_3CHO (ethanal). Chemical reaction shows that CHO and CH_3CO groups are present. The only aldehyde with $M_r = 44$ is CH_3CHO. The fragments from the mass spectrometer trace confirm this.

h Spike at 1700 cm⁻¹ suggests an aldehyde or ketone. Small spikes around 3000 cm⁻¹ suggests CH_2–H group. Lack of broadband around 3200 cm⁻¹ suggests that oxygen is not hydrogen bonded.

i CH_3CH_2OH